NGO Discourses in the Debate on Genetically Modified Crops

The development and use of genetically modified organisms (GMOs) has been a contentious topic for the last three decades. While there have been a number of social science analyses of the issues, this is the first book to assess the role of non-governmental organisations (NGOs) in the debate at such a wide geographic scale.

The various positions, for and against GMOs, particularly with regard to transgenic crops, articulated by NGOs in the debate are dissected, classified and juxtaposed to corresponding campaigns. These are discussed in the context of key conceptual paradigms, including nature fundamentalism and the organic movement, post-colonialism, food sovereignty, anti-globalisation, sustainability and feminism. The book also analyses how NGOs interpret the debate and the persuasive communication tactics they use. This provides greater understanding of the complexity of negotiations in the debate and explains its specific features such as its global scope and difficulty in finding compromises.

The author assesses the long-term interests of various participants and changes in perceptions of science and in public communication as a result. Examples of major NGOs such as Greenpeace, Oxfam and WWF are included, but the author also provides new research into the role of NGOs in Russia.

Ksenia Gerasimova is a research associate and affiliated lecturer in the Centre of Development Studies, University of Cambridge and a research fellow in the Centre of Environment, Energy and Natural Resource Management, University of Cambridge, UK. She is also Professor of Public Policy at the Higher School of Economics University, Moscow, Russia.

Routledge Explorations in Environmental Studies

NGO Discourses in the Debate on Genetically Modified Crops

Ksenia Gerasimova

Routledge
Taylor & Francis Group

LONDON AND NEW YORK

from Routledge

First published 2018 by Routledge

2 Park Square, Milton Park, Abingdon, Oxfordshire OX14 4RN
52 Vanderbilt Avenue, New York, NY 10017

Routledge is an imprint of the Taylor & Francis Group, an informa business

First issued in paperback 2019

British Library Cataloguing-in-Publication Data
A catalogue record for this book is available from the British Library

Library of Congress Cataloging-in-Publication Data
Names: Gerasimova, Ksenia, author.
Title: NGO discourses in the debate on genetically modified crops / Ksenia Gerasimova.
Other titles: Non-governmental organization discourses in the debate on genetically modified crops
Description: Abingdon, Oxon ; New York, NY : Routledge, 2017. | Series: Routledge explorations in environmental studies | Includes bibliographical references.
Identifiers: LCCN 2017002800| ISBN 9781138223899 (hbk) | ISBN 9781315403502 (ebk)
Subjects: LCSH: Transgenic plants—Political aspects. | Transgenic plants—Social aspects. | Transgenic plants—Risk assessment. | Non-governmental organizations.
Classification: LCC SB123.57 .G489 2017 | DDC 631.5/233—dc23
LC record available at https://lccn.loc.gov/2017002800

ISBN: 978-1-138-22389-9 (hbk)
ISBN: 978-0-367-35133-5 (pbk)

Typeset in Goudy
by Apex CoVantage, LLC

This book is dedicated to Miss Jean Ila Currie and her late aunt Miss Edith Whetham, the two women who inspired me to develop my interest in agriculture.

Contents

Tables

Acknowledgements

This research project has been developed under the essential support from the Cambridge Malaysian Commonwealth Studies Centre, the Cambridge China Development Trust, the Cambridge Malaysian Education and Development Trust and the John Templeton Foundation. I express my sincere gratitude to Dr Anil Seal, Fellow of Trinity College, University of Cambridge, Director of Cambridge Malaysian Education and Development Trust (CMEDT), Malaysian Commonwealth Studies Centre (MCSC), Sir Brian Heap, St. Edmund's College, University of Cambridge, Professor Peter Nolan, Founding Director of the Centre of Development Studies, University of Cambridge, and the Director of the project 'The Role of NGOs in the biotech policy-making for food security and climate change' Dr David Bennett, Senior Member, St Edmund's College, Cambridge UK. Thank you!

The Global Food Security Strategic Initiative, University of Cambridge, has served as an important foundation for successful implementation of my research, and I express my gratitude to the Chairs Professor Howard Griffiths and Professor Chris Gilligan and the Coordinator Ms Jacqueline Garget. The Isaac Newton Trust has kindly provided necessary support at a later stage of the project.

I would like to acknowledge the funding support received from the DAAD-University of Cambridge Research Hub for German Studies with funds from the German Federal Foreign Office (FFO).

Many thanks go to Mrs Suzy Adhock, Mr Christian Theil, and Mrs Doreen Woolfrey and to all my students in MPhil in Development Studies who took Paper 600 and contributed to our discussions on genetically modified agriculture. I also thank my proofreader Mrs Sandra Boyd.

Preface

In 2012, I was searching for a new research project after reaching the final stage of my doctoral studies. My main research interest has always been NGOs. A conversation with Cambridge-based plant scientists introduced a new subject – the involvement of NGOs in the debates on genetically modified (GM) crops. As with most members of the general public, those who are not biologists, I had a passing knowledge of genetic manipulation and the first image, to be honest, that would pop up in my mind would have been Mary Shelley's Frankenstein. By the end of 2013, I had embarked on a new research project, enrolling in the University of Cambridge as Research Associate in Social and Biological Studies, which by no means should suggest that I am a biologist. In reality, there was neither monster nor crazy scientists but rather a lot of interviews, talks, observations and reflections.

This project has been a very exciting and unexpected journey. The process of research itself, which includes different interactions, deserves special attention. A second thread in the research has emerged: a study of what it takes to conduct research on GM crops. In the following chapters, I have included these episodes which tell my experience of studying the views of NGOs on GM crops.

1 Introduction

In this introductory chapter, I will discuss the nature of NGOs and why debates on GM crops are such a contested subject.

What are NGOs?

Every work on non-governmental organisations (NGO) starts with a search for their definition and the usual answer is that it is complicated. The Indian parable about an elephant and six blind men, who were able to grasp only one feature of a complex system, is a very suitable analogy for the challenge in defining NGOs. A wide range of synonyms, such as a non-profit organisation, charity or civil society organisation, suggests that these organisations have many characteristics. A GONGO – or government-organised NGO – is an oxymoron which is rather common among NGOs. In other words, NGOs cover a wide range of organisations with different characteristics and that is also why, we argue, different ones present different views on the use of GM crops.

To be able to see NGOs as a whole phenomenon and surpass grasping only one of their features, several definitions might be useful. Thus, NGOs are a truly interdisciplinary subject, and most of the Social Sciences have developed concepts that prove to be relevant in the analysis of NGOs. Table 1.1 provides a summary.

To make sense of this array of definitions, some scholars such as Salamon and Anheier (1992, 1997) have suggested the use of the 'structural/operational

Table 1.1 Definitions of the NGOs

Discipline	Definition of NGOs
Sociology	Associations
Economics	Third sector
International Relations	Non-state actor
Development Studies	Development agency
Experimental Social Sciences	Network, system

definition' and a method of classification to identify systematic differences among NGOs. Their structural/operational definition includes five criteria that define NGOs. NGOs are 'formally constituted', 'non-governmental in basic structure', 'self-governing', 'non-profit-distributing' and 'voluntary to some meaningful extent' (Salamon and Anheier, 1992, p. 268). It is possible to counter-argue every criterion from this list, as there are GONGOS, profit-making organisations, others that are complex in membership and not formally recognised, particularly in a hostile institutional environment, NGOs. On the contrary, a classification exercise might be useful, since it is less exclusive and allows the researcher to discern different features of these numerous organisations. The International Classification of Nonprofit Organizations (ICNPO) is an example of a comprehensive but imperfect NGO typology.

This classification was elaborated upon by the Johns Hopkins Comparative Nonprofit Sector Project on the basis of material from 12 countries. This classification was a step forward from previous classifications that distinguished only 'public serving' and 'member-serving' (Salamon and Abramson, 1982) to a more comprehensive approach that included the 'civic' side of NGOs. Authors themselves are aware that,

> Certain of the distinctions proposed may be difficult to make in practice. Numerous environmental organizations are principally engaged in advocacy activities, for example. Should they be classified accordingly to their area of activity or the nature of their activity? . . .
>
> Beyond this, the nature of a particular type of organization may vary depending on the stage of political and economic development in a country. For example, associations of doctors and lawyers that would be treated as member-serving trade or professional organizations in most developed countries often function as significant promoters of free speech and human rights in developing countries. Unfortunately, the ICNPO system does not take this into account.
>
> (Salamon and Anheier, 1992, p. 284)

The project has identified 11 broad groups of activities in which NGOs get involved (culture and recreation; education and research; health; social services; environment; development and housing; law, advocacy and politics; philanthropy and voluntary activities; international activities, religion, business and professional activities), and one extra group was reserved as not otherwise classified (Salamon and Anheier, 1997, p. 40). It is also true that one organisation can participate in a number of activities named in the list. This is particularly relevant when NGOs take up the case of a complex issue, such as GM crops, which includes education and research, health, social services, the environment, development, lobbying, international activism and culture. The subject that can be agreed upon with the ICNPO relates to the important role played by the institutional context in allowing NGOs to conduct their activities. The level of socio-economic development, the type of political regime and the state of civil society determine what activities NGOs can do and how well.

Almost every modern author working in a subject area has acknowledged the rise in the NGO numbers. Feld and Skjelsbaek recorded the speed at which the sector has grown since the last half of the twentieth century. From around 500 organisations in the 1960s, mainly located in Europe and North America, they increased to the enormous number of thousands and, indeed, millions of organisations internationally (Feld, 1972; Skjelsbaek 1971). Since then, NGOs have been recognised as remarkable actors in international and national political arenas.

Different disciplines have offered explanations as to the rapid growth in numbers and the rising influence of NGOs in international politics. In transnationalism, NGOs are seen as one group of transnational actors that shape transnational networks, a part of a bigger process of globalisation that has been captured in such concepts as global governance and global public space or 'global agora' (Rosenau, 1992; Stone, 2008). There is also a more recent reference linked to these new public arenas (Della Porta et al., 1999).

On the one hand, many issues such as environmental degradation and human rights have become international and interdependent, that is, one actor cannot solve these on its own, and require international cooperation. On the other, transnational networks have become axes of struggle for dominance, leading to asymmetry in power and institutional development (Lennox, 2008). With the example of internationalisation of human rights regimes, Risse and Sikkink showed that both national and international NGOs can be used to circumvent the dominant role of the state and have become very handy in circulating new norms and ideas in surpassing opposition to new principled ideas through tactical concessions, strategic bargaining, moral consciousness-raising and persuasion (Risse and Sikkink, 1999).

While NGOs cannot oppose states on a one-to-one basis, they can appeal to the global polity which include international nongovernmental organisations (INGOs), international organisations (IOs) and other states which can jointly exercise pressure on the state, often successfully. Thus, in this new model called the 'boomerang model', NGOs have become a tool for pressure on states 'from below' (at domestic level) and 'from above' (via transnational networks). It is also clear from the boomerang model that the core of NGO influence is based upon the ideals and ideas (ideology) they propose and promote and how much support can be received for those ideas from other actors.

There also tends to be a geographical pattern: most NGOs are associated with the global North and the global South with social movements (Bendana, 2006). There is also the practice of northern NGOs extracting ideas from the global South and sending it back to southern NGOs 'repacked' as their own (Ludin, 2003). The exchange of ideas and creation of knowledge takes place through networks, as a part of the social movement process.

NGOs, social movements and networks

The relationship of NGOs to networks and social movements is an important question. NGOs are arguably part of the social network. Social networks are described as social movements which bring together different groups and organisations (i.e. NGOs) with various levels of formalisation and interaction among one another (Diani and McAdam, 2003). Both have a target to bring about social and ideological change.

In a social movement process, organisations and individuals are brought together by common goals which then define their strategies for collective action. And although they interact, these interactions are a combination of coordinated and independent actions (Della Porta and Diani, 2006). Through networks, which can then be described as platforms for co-creating ideologies, NGOs interact with one another and participate in social movements. Today, most networks are international which allows them to form 'transnational collective action' (Della Porta and Tarrow, 2005), a set of actions, coordinated along the whole network in the form of campaigns against international organisation, business and states. Three broad processes are found in transnational politics: diffusion ('spread of movement ideas, practices, and frames from one country to another'), domestication or internalisation (external conflicts are mounted at national level) and externalisation – international actors intervening in internal conflicts (Della Porta and Tarrow, 2005, pp. 2–6).

Through transnational collective actions, NGOs have become powerful actors that are potent in bringing three types of change: cognitive, relational and environmental (Della Porta and Tarrow, 2005). Cognitive changes refer to the common adaptation of ideas and strategies leading to formation of newly coordinated and internally negotiated transnational identities and relational changes explain new power dynamics. Environmental change refers to the institutional context in which NGOs operate. It is explained earlier in the boomerang model.

The difference between NGOs and social movements is in the degree of institutionalisation. NGOs are more organised, registered organisations with paid professional staff, while movements are amorphous entities whose members can get together on a particular event and then part. Often as a movement evolves, it produces an NGO. The NGO then professionalises, and issues relating to transparency and accountability appear. Movement members are aware of that and may intentionally choose to opt for civic networks rather than NGOs, and NGO leaders have accepted such restructuring (Bendell and Ellersiek, 2009). This phenomenon has been referred to by activists as the 'movement of movements' (Mertes, 2004).

New networks, these 'movement of movements', have also attracted a new generation of activists who are members of several networks. Such networks may compete and ally with each other for audiences and resources for some campaigns and cooperate in others. This allows high levels of strategic flexibility and dynamism in the interaction between different networks. Tarrow and Della Porta have offered a new vocabulary to describe this phenomenon:

- by 'rooted cosmopolitans', we mean people and groups who are rooted in specific national contexts but who engage in regular activities that require their involvement in transnational networks of contacts and conflicts;

- by 'multiple belongings', we refer to the presence of activists with overlapping membership linked within loosely structured, polycentric networks;
- by 'flexible identities', we mean identities 'characterized by inclusiveness and a positive emphasis upon diversity and cross-fertilization, with limited identification' that 'develop especially around common campaigns on objects perceived as "concrete" and nurtured by an "evangelical" search for dialogue.'

(Tarrow and Della Porta, 2005, p. 237)

Both NGOs and social movements run programmes and campaigns, these being their tools of influence and the execution of ideas. It is difficult to evaluate success of such actions, as there is no unified set of indicators, and there are different types of campaigns. Some are short term, some action driven, others are about long-term impact. As most NGOs receive external funding to run, they must then prove to their donors how successful their performance is; this comes in contradiction to the reference of NGOs to their constituencies as the ultimate authority (Bendana, 2006).

In this regard, it makes sense to briefly discuss who forms NGOs, how they are connected to civil society and how democratic is that process.

Anyone can form an NGO, and such organisations can be a group of just a few people or a huge international organisation with thousands of employees and millions of supporters. For example, in 1979, David McTaggart convinced half a dozen loosely connected early groups to join in a single organisation called Greenpeace. Today, with headquarters in Amsterdam, Greenpeace has gained 2.8 million supporters worldwide and has opened national/regional offices in 41 countries (Greenpeace, 2009). But does the number of necessary global supporters mean that Greenpeace is really representing the views of all its supporters? Or perhaps all these people are reached by different channels, mostly online, and influenced by Greenpeace's formation of ideas.

The fourth position model by Fowler has taken NGOs out of the third sector position (civil society), proposing they be looked at as negotiators and validators located among three sectors: state, business and civil society (Fowler, 2002). Thus, NGOs have a great quality of connectivity, connecting individuals who come together to work on a common cause; they reach out to other groups and movements, organising civil society. They also reach out to state and business through either cooperation or opposition, implemented in the form of campaigns. Generally, NGOs are part of civil society despite moving beyond it in the way their actions are directed.

They are 'a body of individuals who associate for any of three purposes: (1) to perform public tasks that have been delegated to them by the state; (2) to perform public tasks for which there is a demand that neither the state nor for-profit organizations are willing to fulfill; or (3) to influence the direction of policy in the state, the for-profit sector, or other non-profit organizations.'

(Hall, 1987, p. 3)

To reiterate, the main tool of NGOs is ideology and action that support promoted ideas. One of the conceptual frameworks to describe it is the social capital concept. Social capital is generally understood as being connections between individuals, and the common civic values that influence society, and the nature, extent and impact of these interactions (Putnam, 2000; Coleman, 1988).

Putnam argued that these networks 'foster sturdy norms of generalized reciprocity and encourage the emergence of social trust'. When economic and political negotiation is embedded in dense networks of social interaction, incentives for opportunism are reduced (Putnam, 1995, p. 67). Thus, individuals form groups and networks which accumulate trust and are able to overcome the 'free rider' problem and, in that sense, enhance the efficiency of the social interaction and a grassroots democracy. It could have been expected that this would be reciprocated at global level. Yet this process is not straightforward. While Americans and other global citizens meet less to play games, their communication has not ceased; it has simply been transferred online and become global. The intense activity of social media that is seen today suggests that social capital has been created and accumulated online.

Anheier and Kendall (2000) posed a question about the connection of trust, social capital and non-profit organisations. They have distinguished three approaches to the study of trust based on economics, sociology and political science. Interestingly, these three approaches can be relevant to three sectors (business, state and civil society). For example, in market transactions, trust is 'an efficient mechanism to economize transaction costs'. In the social order, trust is a socially constructed reliability, and in social networks, trust is a social capital, a 'civic virtue' (Anheier and Kendall, 2000, p. 347).

Trust in networks becomes a tricky matter. On the one hand, within smaller networks, members of the same network have more interaction; similar backgrounds, interests and goals; and opportunities to build trust. These operate as safety nets for members and word of mouth is highly trusted. In larger networks, those across the globe, trust becomes more complicated. In order to attract members into their network, the organisers of global networks must choose a global cause – an issue that is universal or at least of interest and relevance to many – communicate it in an easily comprehensible manner and also bring emotion to the subject, so that potential members do not choose membership on the basis of rational choice logic or question the achievability of the network's goals; sometimes it may be enough to have high ideals. Environmental issues are a perfect subject on the basis of these criteria.

Conversely, smaller networks may choose neither to reach out to a larger audience nor create a large database of followers and partners; they can simply opt instead for a more closed model of membership and remain as a professional clique.

The increasing number of reported cases of malfeasance by NGOs has raised the question of trust in NGOs and their credibility in general (Gourevitch and Lake, 2012). At global level, NGOs have become serious participants in the political process and have gained the role of ideological arbitrage. But while the classical participants – political parties prove their authority by participating in elections by obtaining public support, that is, they have a democratic mandate to offer their ideas, NGOs involved in political activity do not have to prove their right, yet can suddenly became 'equal to vox populi' (Narochnizkaia, 2008). Calls for accountability and suggestions to clarify and legitimise transparency are hardly new and have long been present in the academic literature and guidebooks for NGOs (Pratt, 2009). But what is in the guidebook and what occurs in reality is a different matter.

To conclude, the major definitional question about NGOs remains the same and unresolved. What NGOs can do is still unclear. In the period 1980 to the present, the world has borne witness to an unprecedented experiment of transferring institutional responsibilities from the state to NGOs. Shockingly, they are not fully understood, since most authors acknowledge the absence of a comprehensive NGO definition (Salamon and Anheier, 1997). Without a definition, they continue to hold the imposed, assumed but not proven, role of real change-bearer. The only part we know is that NGOs, together with other groups and individuals, form social movements and often communicate their ideas via social networks.

GM plants: why plants?

Now we move on to the second object of the study: genetically modified crops, a bone of contention in the studied debates among NGOs.

Genetically modified crops are plants that have genes modified in the laboratory. Genetic modification is 'a technique where individual genes can be copied and transferred to another living organism to alter its genetic makeup and thus incorporate or delete specific characteristics into or from the organism' (Bainbridge et al., 2000). The definitions within genetic engineering on organisms include host or recipient, the microorganism to be modified; the vector or carrier of the new genetic information; insert, or the new genetic information (nucleus acid); donor, the source of new genetic material; and resultant GMO (host(+vector)+insert (Bainbridge et al., 2000).

Manipulations with genes of humans, animals and plants have been met with a certain degree of suspicion and opposition based upon a number of arguments, socio-economic and ethical. While the experiments with animal DNA have raised ethical concerns, they are not comparable with the strength of opposition received by transgenic plants. There seemed to be little opposition to the creation of oncomouse, a mouse used in research to find a cure for cancer (Harraway, 1997). It is an interesting case of the asymmetry under examination:

> The technologies used to produce GM crops are about the same as those for many human drugs including human insulin, and many of the companies involved are the same. It's true that most drug biotechnology involves microorganisms and not higher plants or animals. However, a strain of genetically-modified chickens was recently approved for human drug production – with scarcely a peep from the anti-GMO lobby.
>
> (Daynard, 2015)

There is something special about plants and their genes – as a young geneticist said, 'plants are awesome' (Geisler, 2016). And, indeed, they are special. Plants provide oxygen and food, both essential elements to human survival, and universal to all. In addition, plants are used for most human activities. It is not an exaggeration to say that 'plants have determined the very course of civilization' (Balick and Cox, 1996). Plants are used in production of food and drink, the construction and maintenance of human shelter, producing essential goods, such as medicine and clothing, and even for aesthetic purposes (Swaminathan and Koleher, 1985).

The triad of immobility, carbohydrate production and diverse biochemistry makes plants far more useful to human beings than animals are (Balick and Cox, 1996). Plants not only provide raw materials for production, but they are also used for fuel, either as wood logs or biofuels:

> Though their origins, indeed existence, may seem peripheral to our modern lives – modified or disguised as they often are – plant products are as strategic as oil. In the coming decades, as fossil oil reserves diminish, this importance must surely rise.
>
> (Lewington, 2003, p. 6)

Plants are needed to produce proteins, both vegetarian and animal. So, food is produced from crops which are grown through agricultural practices. Production and consumption of food may vary from one cultural environment to another, but they are a set of processes common to every human and are central to human interaction.

First, food lies at the heart of social activity. How many of us meet over lunch or for coffee? Second, food is political:

> The physiological need in humans to eat every day makes access to food a crucial issue . . . It also makes us vulnerable, weak and easy to control. In this way food is entrenched in structure of subordination, governance and domination.
>
> (Lien, 2004, p. 6)

Finally, plants have a wide variety of genes. For rice, for example, the number of genes is around 41,000 (Sterck et al., 2007). The combination of their essential status to food and the diversity of their genetic material makes plants important and can raise concerns over genetic manipulation. The first question that can be posed is why do we need GM crops at all? Farmers grow plants to feed themselves and their families and to sell crops for a profit. Agrobusinesses also grow crops for profit. Agricultural production has excelled over times, yet the leading factors in crop losses are natural disasters, including floods and drought, and natural organisms, such as weeds, pests and viruses. As pointed out by Brandon Mitchener, the Monsanto public affairs head in Europe, the Mideast and Africa:

> What all farmers want is to produce more with less. They have a finite amount of land, they have a finite amount of resources, and they want to

get the maximum yield from their land with the seed and the water and the manpower they have to farm that land.

(Shoo, 2014)

The same point is illustrated by the list of commercial GM traits: Abiotic Stress Tolerance, Altered Growth/Yield, Disease Resistance, Herbicide Tolerance, Insect Resistance, Modified Product Quality and Pollination control system, all of which serve to improve quantity and quality of crops (ISAAA, 2016).

Thus, if science has a new solution to these old problems, both farmers and agribusiness would be willing to try a new technique or new seeds. Another commonality is that scientists, agribusiness and farmers are focused on results: 'For a farmer or a geneticist, we use whatever tool will work' (Garthwaithe, 2014). The existing GM plants are those that are grown for commercial agriculture. Today's GM plants include alfalfa, apple, Argentine canola, bean, carnation, chicory, cotton, creeping bentgrass, eggplant, eucalyptus, flax, maize, melon, papaya, petunia, plum, polish canola, poplar, potato, rice, rose, soybean, squash, sugar beet, sugarcane, sweet pepper, tobacco, tomato and wheat (ISAAA, 2016).

The argument that GM seeds might yield anticipated harvests of crops, at least on a short-term basis, can be illustrated with the statistics on GM crops spread throughout the world. ISAAA is an NGO which shares recent data on genetic engineering in crops and lists the following countries as having approved GM events: Argentina, Australia, Bangladesh, Bolivia, Brazil, Burkina Faso, Canada, Chile, China, Colombia, Costa Rica, Cuba, Egypt, the European Union (EU), Honduras, India, Indonesia, Iran, Japan, Malaysia, Mexico, Myanmar, New Zealand, Norway, Pakistan, Panama, Paraguay, Philippines, the Russian Federation, Singapore, South Africa, South Korea, Sudan, Switzerland, Taiwan, Thailand, Turkey, the United States, Uruguay, Vietnam (ISAAA, 2016).

Concerned scientists have argued that GM crops are not harmful on their own, but statistically with the rise in numbers and varieties, risks can increase. The risk from GM plants may be similar to those resulting from the introduction of non-native organisms, which in the majority cannot survive although a few may thrive and cause damage. As the ecosystem dynamics are not fully understood, GM crops might have unknown risks in the future (Rissler and Mellon, 1996). There may be both biological and socio-economic risks, and these are intertwined. From a biological perspective, transgenic crops may become weeds or they could move genes to weed, develop new vital pathogens and thus threaten global centres of crop diversity (Rissler and Mellon, 1996, p. xi). From the socio-economic perspective, it is argued that commercialisation of transgenic crops could result in releasing GMOs under uncontrolled conditions, which could, in turn, lead to the development of illicit, dangerous recreational addictive drugs (Geber and Young, 1993). Most of the risks, however, are argued to be just potential; it is recommended that GM crop risk assessment be undertaken on a case-by-case basis (Garthwaithe, 2014). The absence of final risk assessment results opens up a space for debates with ensuing claims that are difficult to prove or disprove.

It is also difficult to create a co-existence model which would allow GM crops to be grown only in a controlled territory. There have been cases of non-GM crop contamination with transgenic material. One of the most famous cases is Monsanto versus Schmeiser which took place in 2004. Percy Schmeiser, a Canadian farmer, constituted a patent infringement against Monsanto because his canola harvest contained genetic traits patented by Monsanto, and this genetic contamination might have been accidental, driven either by pollen or wind-borne seeds (Kinchy, 2012).

Another bone of contention about GM plants is their 'unnatural' nature. Although they are organic entities and biological objects, their genomes have been modified in a lab, thus they are not products of nature, but 'authentic products of technological science', and that raises ethical concerns and fears (Lacey, 2005).

NGOs in the debate on GM crops and how to study them

The brief review of risks associated with GM crops, given earlier, has shown that there are several bones of contention, and one big debate (Debate), in fact, falls into many interlinked smaller discussions: on human health risks, ecological consequences, land use, seeds ownership, regional food security, the interests of farmers and big agri-corporations and, finally, ethical issues and views on life. Each of these debates then spreads into smaller conversations around concrete case studies and narrower questions. In fact, many general public discussions of GM crops have been held in the form of debates. It has become a good tradition for a public event to have a debate over the use of GM crops by inviting representatives of both sides.

The application of GM crops has been studied globally (Qaim, 2016) and regionally, for example, GM crops in Africa (Thomson, 2003); by crop, such as Bt cotton, GM maize and soybean (Mohamed, 2015; Scoones, 2003); from the point of risk assessment (Hull et al., 2009), regulation (Graham, 2000), by certain risk groupings such as environmental, social and ethical (Wu, 2000; van Emden and Gray, 2004); by actors, such as farmers, NGOs and agribusiness (Schurman and Munro, 2010; Sajeev, 2012; Robin, 2008); and, finally, as a public debate which is also studied by region or by country (Masood, 2005; Horlick-Jones et al., 2007). Many articles and books are written by activists such as Claire Robinson, Michael Antoniou and Vandana Shiva, to name a few, and are, therefore, in this research considered as a source of primary data.

Studying NGOs in general is a sensitive matter, and this most definitely applies to this research question relating to NGO participation in debates over GM crops. As shown earlier, the researchers do not often know what a NGO is nor how many are active, so most existing studies of NGOs are qualitative in-depth studies, and quantitative methodology may well be difficult to apply to this subject.

Perhaps with some degree of the idealism of the young researcher, I have intended to conduct my research in the most comprehensive, neutral manner. To my surprise, if I may say so, radical representatives from both sides of the debates did not like the idea. The general message was that I should make up my mind and advocate for one 'right' side of the debate; talking to the other side was not worth it.

It has turned out that with this research I have stepped into a sensitive matter. One NGO that I approached, the Soil Association, has asked not to be included in this research. At some points, I have almost been bullied. For the conference 'Past, Present and Future of Europe 2014' I submitted a piece entitled 'Why are Europeans against GM crops?' for the panel concerned with the changing role of science. The abstract was accepted, and the chair wrote to inform that he was on the other side of the debate and looked forward to debating with me publicly. After explaining that I was not interested in a public fight, the chair withdrew his decision and suggested that I transfer to another panel; I did so and went on to present my paper (Gerasimova, 2016).

Experiencing this has not deterred me from conducting research on NGOs in the future; in fact, I have decided to take a step back and, unlike most existing research, will not form my own views upon the subject (GM crops). I plan rather to study the views of NGO members on the subject, classify their ideas, and do not claim comprehensiveness in my research. For these purposes, social anthropology has proven to be very useful. Thus, my discourse analysis, unlike the Foucaultian one, is not 'freed from the anthropological theme' (Foucault, 1972, p. 16).

The socio-anthropological approach to debates on genetically modified crops and the focus on debate topics puts the author at a distance from the political heat of the debate and enables a cool, balanced viewpoint. The analytical framework used is the emic approach which comes from within the social group and the perspective of the subject, presenting the views of participants within the debate in a culturally relative manner – that is, by avoiding making judgements.

The emic methodology uncovers a wide range of socio-anthropological issues discussed in the debate: problems that seem to be philosophical, such as views on nature and its meaning to humans, spiritual existence and moral ideas, epistemo- logical questions about knowledge and science and patterns of social life such as the production and circulation of food, gender roles and public communication.

Also in the beginning, I attempted to compile my own database of the NGOs participating in the debate. With the existing literature and checking the internet for contact details, I put together a preliminary list of 358 non-governmental organisations from the studied regions participating in the current debate of GM crops and seven movements which could have been a representative sample of all NGOs participating in the debate. However, there is no official data on the numbers of NGOs (the total studied entity) generally, making it impossible to estimate the exact numbers of the studied entity to run a large-scale random- sampling survey. Under such circumstances, and in order to show the different types, the method of classification, as discussed earlier (Salamon and Anheier, 1997) was chosen.

Here, I encountered the classic problem of classifying NGOs and social movements – how to classify environmental NGOs which are numerous but not a unified community. They vary in their understandings of how to promote the protection of the natural environment, and often the boundaries between environmentalism and other NGO activities are blurred (Ahmed and Potter, 2006, p. 211). LobbyWatch compiled a database of 'deceptive PR', which includes 'front groups/ lobby groups/ think tanks', 'neo-conservative lobbyists', 'living Marxism links', 'fake persuaders', 'third world lobbyists', 'PR operators', 'industry lobby group', 'corporate science' and 'industry friendly experts'. This classification has mixed individuals and state organisations (USAID) and NGOs in the same categories. Altogether, the data set lists 62 organisations and 52 individuals (LobbyWatch, 2016). Genetic Literacy Project has conducted research on networking status of Dr Vandana Shiva and her organisation Navdanya. It is argued that she might be the best connected anti-GM activist, and her organisation has eight NGOs/foundations and one individual as funders and 40 NGOs and one individual – as partners (Genetic Literacy Project, 2015).

This study provides the list of NGOs participating in the debate which was divided into eight categories, these being business associations, farmers' associations, foundations, humanitarian NGOs, international think tanks, green/ social and environmental justice NGOs, scientific research and science lobbying organisations, and social movements based on two major criteria such as the constituencies they represent and their organisational structure. Green NGOs often engage in social justice activities; thus, it was sensible not to divide these into two categories but keep them as one. Appendix 1 graphically shows the distribution of NGOs among these seven groups. Green and social justice NGOs are the most numerous groups within the entity.

Understanding that it is impossible to conduct a comprehensive dataset of all NGOs involved in the debate and run a statistical study, I have settled down with the idea that I would investigate only the ideas on GM crops presented by available NGOs and their communications to present these ideas. Six broad discourses, such as nature fundamentalism, colonial discourse, regionalism, sustainability discourse, alterglobalism and feminist discourses have been identified in the debate, although the list of discourses may be extended.

To explain this methodology briefly, 'discourse' refers not just to the language being used in the discussion but to 'the way that language (and beyond) operates to produce meanings' which are 'culturally and historically located' to constitute knowledge, social relations and social identity (Gillies, 2009). Critical discourse analysis (CDA) is a problem-orientated investigation of semiotic data (written, spoken and visual) which is seen as a social practice in order to 'de-mystify ideologies' (Wodak and Meyer, 2009, p. 3). Another method which could be useful is Social Network Analysis (SNA), a relatively new method receiving increasing support from social scientists. It is also used in research on NGOs. It allows for the identification of network structures, alliances and platforms, and in the understanding of organisational hierarchy within the studied network. Examples of such studies are 'Assessing Social Network Structure of Stakeholder Organizations in the Grenadine Islands' by Blackman and Mahon (2013) and 'Agency and Social Networks: Strategies of Action in a Social Structure of Position, Opposition, and Opportunity' by Stevenson and Greenberg (2000). The latter explains the success and failure of actors (NGOs) in a network trying to influence policies on environmental issues in a small city.

Interestingly, members of the NGOs involved in the debate also use this method. I was shown by an African activist a social network diagram which classified pro-GM NGOs, their ties and connections to funding organisations. In this research, it was possible to partially implement this methodology while discussing diffusion and the transfer of ideas and executing impact within a global network of NGOs discussing GM crops (Part 2).

By deciding to study the ideas presented by NGOs in the debate, I have referred to discourse analysis, where discourse in its general sense refers to written and oral communications – in this case, about a particular subject by a particular group of people. Yet I learnt that the terms 'discourse' and 'discourse analysis' are as contested as the subject of my research, and there at least 57 varieties of discourse analysis (Bauer and Gaskell, 2000). It also seemed that existing work on the discourse of GM debates analyse it as being just one or two issues – for and against GMO – while in this book, I distinguish different sets of ideas and ideologies involved which support the argument that one debate falls into several. I have identified different sets of ideas and matched them with the relevant campaigns.

Plan of the book

This book sets out to take the debate on genetically modified crops more deeply and dissect it at the conceptual level to understand the nature of persuasion used. It maps advocacy campaigns led by NGOs and juxtaposes them to the various concepts they represent. This allows us to obtain insights on the consciousness, ideas and other personal and interpersonal experiences of debate participants; how they are brought into public discussion; and how participants choose their argument and positions, convince, call for action and relate to alternative perspectives. It also helps to provide an overview of the concepts dominating the debate. The book is structured in two parts: the first describes the six main discourses identified, while the second analyses the debates through an examination of the communicative reasoning processes – the phenomenon of advocacy science, reasons for changing arguments and sides and global reciprocation within the debates and transfer of ideas. The concluding chapter will provide an overview of how the debate is framed and discusses the communication strategies used.

Literature

Ahmed, S., Potter, D.M. (2006) NGOs in international politics. Bloomfield: Kumarian Press.
Anheier, H., Kendall, J. (2000) Interpersonal trust and voluntary associations: Examining three approaches. The British Journal of Sociology, 53(3): 343–362.
Bainbridge, J., Ellahi, B., Smith, G., Whisson, J. (eds.) (2000) Genetically modified foods: A practical guide for business. Oxford: Chandos.
Balick, M.J., Cox, P.A. (1996) Plants, people and culture: The science of ethnobotany. New York: Scientific American Library.
Bauer M.W., Gaskell, G. (2000) Qualitative researching with text, image and sound. London: Sage.
Bendana, A. (2006) NGOs and social movements a North/South divide? Civil Society and Social Movements Paper N22 June 2006. Geneve: UNRISD.

Bendell, J., Ellersiek, A. (2009) *Noble networks? Advocacy for global justice and the 'network effect'*. Civil Society and Social Movements Paper N31. Geneve: UNRISD.

Blackman, K., Mahon, R. (2013) *Comparison of social networking among organisations in the Grenadine Islands between 2005 and 2010*. CERMES Technical Report No 66. University of the West Indies: Cave Hill.

Coleman, J.S. (1988) Social capital in the creation of human capital. *American Journal of Sociology*, 94: 95–120.

Daynard, T. (2015) Agricultural anti-GMO activism is probably not about the technology. http://tdaynard.com/2015/12/29/agricultural-anti-gmo-activism-is-probably-not-about-the-technology-at-all/

Della Porta, D., Diani, M. (2006) *Social movement: An introduction*. Blackwell: Oxford.

Della Porta, D., Kriesi, H., Rucht, D. (1999) *Social movements in a globalizing world*. Basingstoke: Macmillan.

Della Porta, D., Tarrow, S. (2005) Transnational processes and social activism: An introduction, in D. Della Porta, S. Tarrow (eds.) *Transnational protest and global activism: People, passions and power*. Lanham: Rowman & Littlefield, pp. 1–17.

Diani, M., McAdam, D. (2003) *Social movements and networks: Relational approaches to collective action*. Oxford: Oxford University Press.

Feld, W.J. (1972) *Nongovernmental forces and world politics: A study of business, labor, and political groups*. New York: Praeger Publishers.

Foucault, M. (1972) *Archaelogy of knowledge*. London: Tavistock Publications.

Fowler, A. (2002) NGO futures: Beyond aid: NGDO values and the fourth position, in M. Edwards, A. Fowler (eds.) *The Earthscan reader on NGO management*. London: Earthscan, pp. 13–26.

Garthwaithe, J. (2014) Beyond GMOs: The rise of synthetic biology. The Atlantic, 24 September 2014. www.theatlantic.com/technology/archive/2014/beyond-gmos-the-rise-of-synthetic-biology/380770 as viewed 29.03.2016.

Geber, J., Young, A. (1993) Social implications of transgenic plants, in S.D. Kung, R. Wu (eds.) *Transgenic plants vol. 2 present status and social and economic impacts*. London: Academic Press, pp. 217–228.

Geisler, K. (2016) The 5 minute introductory challenge talk, FarmRound 2016, Queens College, Cambridge, 5 July 2016.

Genetic Literacy Project (2015) Biotech gallery: Vandana Shiva: 'Rock Star' of GMO movement has anti-science history. www.geneticliteracyproject.org/glp-facts/vandana-shiva as viewed 16.07.2016.

Gerasimova, K. (2016) Genetically modified crops: Why Europe is against?, in Yolanda Espiña (ed.) *Images of Europe past, present and future*. Porto: Universidade Catolica Editora, pp. 1034–1043.

Gillies, D. (2009) *Critical discourse analysis and current education policy*. Strathclyde: University of Strathclyde.

Gourevitch, P.A., Lake, D.A. (2012) Beyond virtue: Evaluating and enhancing the credibility of non-governmental organizations, in P.A. Gourevitch, D.A. Lake, J.G. Stein (eds.) *The credibility of transnational NGOs: When virtue is not enough*. New York: Cambridge University Press, pp. 3–33.

Graham, V. (2000) *The EU and genetically modified foods: Current regulations and future trends (Chandos series on the food industry)*. Oxford: Chandos Publishing.

Greenpeace (2009) Questions about Greenpeace in general. *Background*, 8 January 2009. www.greenpeace.org/international/en/about/faq_old/questions-about-greenpeace-in/

Hall, P. (1987) A historical overview of the private nonprofit sector, in W. Powell (ed.) *The nonprofit sector: A research handbook*. New Haven: Yale University Press, pp. 3–26.

Harraway, D. (1997) *Modest-Witness@Second-Millenium.FemaleMan-Meets-OncoMouse: Feminism and technoscience.* New York: Routledge.

Horlick-Jones, T. (2009) *The GM debate: Risk, politics and public engagement.* London: Routledge.

Hull, R., Tzotzos, G., Head, G. (2009) *Genetically modified Plants. Assessing safety and managing risks.* London: Academic Press.

ISAAA (2016) GM approval database. www.isaaa.org/gmapprovaldatabase/ as viewed 29. 03.2016.

Kinchy, A. (2012) *Seeds, science and struggle: The global politics of transgenic crops.* Cambridge: MIT Press.

Lacey, H. (2005) *Values and objectivity in science the current controversy about transgenic crops.* Lexington: Rowman & Littlefield.

Lennox, V. (2008) Conceptualising global governance in international relations. E-International Relations Students. 3 October 2008. www.e-ir.info/2008/10/03/ conceptualising-global-governance-in-international-relations/ as viewed 23.03.2016.

Lewington, A. (2003) *Plants for people.* London: Eden Project Books.

Lien, M.E. (2004) Politics of Food, in M.E. Lien, B. Nerlich (eds.) *Politics of Food.* Oxford: Berg.

LobbyWatch (2016) WHO'S who among the lobbyists www.lobbywatch.org/lm_profiles. html as viewed 06.07.2016.

Ludin, J. (2003) Where are we with North-South learning? www.bond.org.uk/networker/2002/ dec02/opinion.htm as viewed 29.03.2016.

Masood, E. (2005) *The GM debate – who decides?: An analysis of decision-making about genetically modified crops in developing countries.* London: Panos Publications.

Mertes, T. (2004) *A movement of movements.* New York: Verso.

Mohamed, E. (2015) *Genetically modified soybean and maize: Biosafety assessment.* London: Scholars Press.

Narochnizkaia, N.A. (2008) Demokratia XXI veka: Pererozdenie smyslov I zennostey, in N. Narochnizkaya (ed.) *Oranzevye Seti. Ot Belgrada do Bishkeka.* Moscow: Alteya, pp. 5–10.

Pratt, B. (2009) *Legitimacy and transparency for NGOs.* Oxford: INTRAC.

Putnam, R. (1995) Bowling alone: Democracy in America at the end of the twentieth century. *Journal of Democracy,* 6(1): 65–78.

Putnam, R. (2000) *Bowling alone.* New York: Simon & Schuster.

Qaim, M. (2016) *Genetically modified crops and agricultural development.* London: Palgrave MacMillan.

Risse, T., Sikkink, K. (1999) The socialization of international human rights norms into domestic practices: Introduction, in T. Risse, S. Ropp, K. Sikkink (eds.) *The power of human rights.* International Norms and Domestic Change. Cambridge: Cambridge University Press, pp. 1–38.

Rissler, J., Mellon, M. (1996) *The ecological risks of engineered crops.* Cambridge: MIT Press.

Robin, M.M. (2008) *The world according to Monsanto: Pollution, corruption, and the control of the world's food supply.* Paris: Découverte; Issy-les-Moulineaux.

Rosenau, J. (1992) Governance, order and change in world politics, in J.N. Rosenau, E.-O. Czempiel (eds.) *Governance without government: Order and change in world politics.* Cambridge: Cambridge University Press, pp. 1–29.

Sajeev, M.V. (2012) *Scientists' perception and farmers' readiness towards GM crops: An Indian perspectives.* Saarbrücken: LAP LAMBERT.

Salamon, L.M., Abramson, A. (1982) *The Federal budget and the nonprofit.* Washington, DC: Urban Institute.

Salamon, L.M., Anheier, H.K. (1992) In search of the non-profit sector II: The problem of classification. *Voluntas*, 3(3): 267–309.

Salamon, L.M., Anheier, H.K. (1997) *Defining the nonprofit sector: A cross-national analysis*. Manchester: Manchester University Press.

Schurman, R., Munro, W.A. (2010) *Fighting for the future of food: Activists versus agribusiness in the struggle over biotechnology*. Minneapolis: Minnesota Press.

Scoones, I. (2003) *Regulatory manoeuvres: The Bt cotton controversy in India*. Brighton: IDS.

Shoo, E. (2014) Can genetically modified crops end hunger in Africa? *Deutsche Welle*. www.dw.com/en/can-genetically-modified-crops-end-hunger-in-africa/a-17385964 as viewed 24.01.2014.

Skjelsbaek, K. (1971) The growth of international nongovernmental organizations in Europe. *International Organization*, 25(3): 420–442.

Sterck, L., Rombauts, S., Vandepoele, K., Rouzé, P., Van de Peer, Y. (2007) How many genes are there in plants (. . . and why are they there)? *Current Opinion in Plant Biology*, 10(2): 199–203. Epub 2007 Feb 7.

Stevenson, W.B., Greenberg, D. (2000) Agency and social networks: Strategies of action in a social structure of position, opposition, and opportunity, *Administrative Science Quarterly*, 45 (4): 651–678.

Stone, D. (2008) Global public policy, transnational policy communities, and their networks. *Policy Studies Journal*, 36(1): 19–38.

Swaminathan, M.S., Koleher, S.L. (1985) *Plants and society*. London: Macmillan.

Tarrow, S., Della Porta, D. (2005) Conclusion: 'Globalization', complex internalism, and transnational contention, in D. Della Porta, S. Tarrow (eds.) *Transnational protest and global activism: People, passions and power*. Lanham: Rowman & Littlefield, pp. 227–274.

Thomson, J.A. (2003) *Genes for Africa: Genetically modified crops in the developing world*. Lansdowne: UCT Press.

Van Emden, H.F., Gray, A.J. (2004) GM *crops: Ecological dimensions*. Aspects of Applied Biology 74. Wellesbourne: Association of Applied Biologists.

Wodak, R., Meyer, M. (2009) *Methods for critical discourse analysis*. London: Sage.

Wu, W.J. (2000) GM *crops ecological risk assessment: The transfer Bt maize*. Beijing: Chemical Industry Press.

Part 1

Discourses in the debate

What is discourse, and how to implement discourse analysis?

Discourse analysis (DA) studies how the use of language represents different views of the world and its different interpretations. It can consider how these different views of the world are constructed through the use of discourse (Paltridge, 2006, p. 2). Another term is 'critical discourse analysis' (CDA).

Teubert distinguished 'discourse at large' and special discourses which are subsets of this discourse. The large discourse which 'consists of all spoken, written or signed utterances from the time when people started using language', cannot be researched in full. What can be researched, however, is a selection of particular discourses and participants' contributions to such discourses (Teubert, 2010, p. 116). With regards to this research, while it is not possible to study the global discourse of NGOs discussing the use of genetically modified crops, it is possible to identify a number of smaller discourses limited either by a particular group or a geographical region.

Each discourse consists of discourse objects, surrounding concepts, observed and observers, who form a discourse community, 'a group of people participating in the same discourse' (Teubert, 2010, p. 51). It can be oral or written. Contribution to discourse presupposes agency and intentionality. In this study, a studied discourse community is members of NGOs and associated social movements who publicly presented their opinions on GM crops.

On a linguistic basis, discourse is constructed by content or propositions. Propositions are useful for understanding the meaning of discourse, as they play three distinct roles: they are 'bearers of truth and falsity and other modalities, such as necessity, possibility, justification', they 'serve as the objects of propositional attitudes, such as belief and desire' and they 'articulate the content expressed by linguistic material' (Collins, 2011, pp. 2–3).

A discourse community shares common goals and channels of communication and exchange information. It also has preferred genres, its own terminology and vocabulary, which can be formally agreed or be tacit (Swales, 1990).

Devitt distinguished three types of language or what he called discourse user groups: communities (group of people sharing time in common activities), collectives (a group focused around a particular interest) and a network, which is a group of people who may or may have not met each other, yet are participating in the same discourse (Devitt, 1997). A network also deals with the dimension of solidarity (Milroy and Milroy, 1997). The latter type – a network – is particularly relevant to the NGOs' involvement in the debate on the use of GM crops. Different organisations, as will be shown in the following six chapters, use different ideas to discuss the subject of GM crops and jointly form a general discourse of NGOs assessing GM crops.

For Durkheim and Teubert, society is not just a sum of individuals and their consciousnesses (Durkheim, 1982; Teubert, 2010). There is an interaction of plural consciousnesses that can be called collaborative cognition. In such social processes, participants of the discourse contribute their utterances and interpret verbal and non-verbal interactions. They negotiate meanings in discourse jointly, as a collaborative act (Teubert, 2010).

According to Longino, discursive interactions are social processes of knowledge production, in which its actors introduce and sustain the information that is considered knowledge (Longino, 2002).

There is a distinction of the traditional epistemological understanding of knowledge as 'justified true belief' from 'knowledge as it is being used by real people in real communities', which is defined in more relative and contextual terms, that is, in the system of beliefs and 'the knowledge standards or criteria of a community' (van Dijk, 2014, p. 16).

Actors of a discourse community decide on the meaning of the discourse: 'Meaning is what is exchanged and shared in contributions to a discourse' (Teubert, 2010, p. 208). Thus, meaning is social, not mental. Meaning can be found within the discourse. It is the collaborative act of interpretation that makes meaning available to the engaged discourse community (Teubert, 2010, p. 208).

Discourse is a social construction and perception of reality which is embedded in social and cultural practices. Members of a discursive community express social identity such as gender, age, location, class, education and profession through discourse. In certain discourses, one identity can become determinant, such as gender for a feminist discourse, for example.

Past experiences and previous ideas influence the creation of discourse. On one hand, new ideas are created, on the other, meaning creation, 'understanding' can be part of participation in an event of tradition, so it is a process of transmission in which past and present are constantly mediated (Gadamer, 1989, p. 291).

Also, since the creation of discourse is a collective action, it is logical to connect the discourse creation of a social group with the collective actions of that group. It is possible to suggest that certain ideas bring people together and they form an organisation and vice versa an existing organisation participates in creating ideas. Thus, members of movements and organisations jointly create ideas from already formed organisations or create new ones and conduct particular public actions, for example, in campaigns to achieve ideological targets, as shown in Figure 2.1.

Figure 2.1 NGOs meeting ideological targets

Another important part in understanding discourse is its context. Context is essential to understanding and interpreting the meaning of discourse – be it physical context or social context. There are also situational and background contexts (Paltridge, 2006, p. 54). For example, developed and developing countries might have different contexts.

There are cross-cultural and linguistic differences – the concepts may be difficult to translate and may not fit into different cultural contexts (Wierzbicka, 2003). In the GM debate, many activists use English, but there are those who use German (Koehlin) and French (Séralini). In a study of a general discourse analyses, their ideas are brought in one language and, as a result, a specific aspect of a particular meaning may be partially lost in translation (into English). Also, certain concepts are expressions of specific needs or values shared by a specific community at a specific time which can either stay embedded (organic agriculture in Germany) or be overturned into an opposite system of values (epigenetics in Germany).

Another interesting question is whether to begin analysing first the discourse material, such as text or a speech, or its context. In a way, it is a chicken-or-egg question. It is difficult to choose the priority. What is evident is these two reflect each other.

Discourse can be created in different genres – in both spoken and written – emails, books, blogs, lectures, conference presentations, peer-reviewed articles, book chapters and so forth. So another useful branch of CDA is genre analysis which includes the analysis of setting of the text (locus); its focus of the text; its purpose of the text; author, audience and their relationship; expectations and conventions; background knowledge; values; and understanding and relationship with the text (Paltridge, 2006, p. 100).

So, altogether, discourse analysis can include identification of major discourse themes (topics); local coherence, that is, how knowledge is organised (what is sequence of propositions); actors' descriptions; levels, details and precision of description; implications and presuppositions; definitions and other means (special discursive moves); evidentiality (or references); argumentation; metaphor (conceptualisation of abstract and complex knowledge); modalities (of events and knowledge); rhetoric devices (hyperboles and euphemisms); grammar; lexicon and nonverbal ('semiotic') statures (picture, graphs, film) (van Dijk, 2011, pp. 37–39).

In 1946, Morris classified discourses according to modes of signifying and the use of sign complexes ('primary sign use') and came up with 23 possible basic types of discourse. These include valuative, incitive, systemic, designative, scientific, fictional, legal, cosmological, appraisive, mythical, poetic, moral, critical, prescriptive, technological, political, religious, propagandistic, formative, logical, rhetorical, grammatical, and metaphysical (Morris, 1946). While this classification might be helpful, it is also a finding of this study that a particular discourse can represent more than one type of the Morris classification. For example, most discourses identified in the NGOs debate over GM crops are valuative, political and critical.

In this work, discourse analysis is done in three steps: identification of actors of the discourse (NGOs), classification of their concepts and ideology and matching them with those NGOs' campaigns that represent a relevant discourse through public actions.

Literature

Collins, J. (2011) *The unity of linguistic meaning.* Oxford University Press: Oxford.

Devitt, A. (1997) Genre as a language standard, in W. Bishop, H. Ostrum (eds.) *Writing genres.* Carbondale: Southern Illinois University Press.

Durkheim, E. (1982) The material conception of history, in S. Lukes (ed.) *The rules of sociological method.* New York: Free Press, pp. 167–174.

Gadamer, H.G. (1989) *Truth and method.* New York: Continuum.

Longino, H.E. (2002) *The fate of knowledge.* Princeton: Princeton University Press.

Milroy, J., Milroy, L. (1997) Varieties and variation, in F. Coulman (ed.) *The handbook of sociolinguistics.* Oxford: Blackwell, pp. 47–64.

Morris, C.W. (1946) *Signs, language and behavior.* New York: Braziller.

Paltridge, B. (2006) *Discourse analysis: An introduction.* London: Continuum.

Swales, J.M. (1990) *Genre analysis: English in academic and research settings.* Cambridge: Cambridge University Press.

Teubert, W. (2010) *Meaning, discourse and society.* Cambridge: Cambridge University Press.

van Dijk, T.A. (2011) Discourse, knowledge, power and politics: Towards critical epistemic discourse analysis, in C. Hart (ed.) *Critical discourse studies in context and cognition: Discourse approaches to politics, society and culture.* Vol. 43. Amsterdam: John Benjamin, pp. 27–63.

van Dijk, T.A. (2014) *Discourse and knowledge: A sociocognitive approach* Cambridge: Cambridge University Press.

Wierzbicka, A. (2003) *Cross-cultural pragmatics: The semantics of human interaction.* New York: Mouton de Gruyter.

2   Nature fundamentalism

Nature fundamentalism: discovering and describing the discourse

An alternative name for this chapter would be mystical environmentalism. My first encounter with this type of movement occurred in Russia in 2004. I was an enthusiastic volunteer and project leader responsible for a dozen foreigners who arrived in Russia to experience life in the Grishino eco-village. The goal of the project was to help residents in their daily life tasks, such as building a sauna house, managing gardens and undertaking household chores. In return, volunteers were accommodated in wooden houses and provided with vegetarian meals. The settlement is located in the abandoned village of Leningradskaya Oblast, north of St. Petersburg. These were a remarkable two weeks. Not only did I manage to stay away from vipers and midges, but I also provided meals for my volunteers which was not easy, given the lifestyle followed by the Grishino residents. Most of their food came from nature – either wild or cultivated in the garden. There was not much to eat to say the least. The story, however, has a happy ending – we all survived, lost a couple of kilos and I also learnt about an eco-village in Scotland called Findhorn. I was told that magic happens there – by speaking to plants with love, local eco-villagers grow crops bigger in size and much tastier than average crops. Once I moved to Cambridge, I began to grow tomatoes that, alas, despite my words of love, have shown good crops though nothing close to the fabled vegetables of Findhorn. In Grishino, we often sang a song 'Earth My Body' which I liked for its catchy tune.

Another useful experience was in 2014 when I visited Berlin to attend the Farbe der Forschung (The Colour of Science) conference in the hopes of speaking to Benny Haerlin who was one of the Greenpeace activists involved in the GM crops debate. Attending the conference and talking to its participants helped me to discover a whole new discourse within the debate, which can be defined as nature fundamentalism or mystical environmentalism. The conference was my first point of contact with the German biodynamic agriculture school of thought. A further analysis of the ideas of contemporary biodynamic supporters reveals a reference to earlier concepts developed by Rudolf Steiner and his followers.

To explain the chosen name for the discourse analysed in this chapter, fundamentalism can be defined as 'a religious movement characterized by a strict belief in the literal interpretation of religious texts', 'strict adherence to any set of basic ideas or principles' or, simply, the beliefs held by those in a religious movement (Thesaurus, 2016). In fact, all three definitions might be relevant in this case.

In his Anniversary Address at the Royal Society in 2002 Lord May, as president, spoke of the values of science being those values introduced during the Enlightenment (cited in Cook, 2004, pp. 32–35). They are rational, humane and questioning. He contrasted these values with the 'fundamentalist belief systems' which include texts, dogma, ideology or revelation and are 'authoritarian, seeking to suppress questioning' (Cook, 2004, pp. 32–35). The two approaches or paradigms clash over the understanding of two very fundamental concepts – nature and science. In Lord May's view, fundamentalists believe that 'truth cannot be reliably established by subjecting hypotheses to soulless experimental tests' and irrationally oppose any change. For example, they propagate 'outdated' nineteenth century agricultural practices, and with 'their traditional knowledge' and 'instinctive beliefs' they distrust new ideas and new technologies, including genetic engineering, and are thus resistant to the idea of GM crops (Cook, 2004, pp. 32–35).

In the discourse described in this chapter, there is a well-developed system of ideas, 'religious texts', and a network of organisations and movements that promote their ideology and run campaigns. In terms of their stand on the GM debate, they are opposed to the use of transgenic plants.

Steiner: ideas, texts and organisations of reference

Rudolf Steiner (1861–1925) was a German philosopher and leader of the anthroposophy movement which summarised a number of ideas from German philosophers such as Christian Rosenkreuz, Kant, Fichte, Schelling and Goethe. There is a widely shared feeling among anthroposophists that his ideas were unique and incomparable (McDermott, 2015). In terms of the movement's ideological input, Steiner was the founder and produced much knowledge on different aspects of life, leaving nearly 400 volumes of material (Lindenberg, 2012). His work can be called spiritual philosophy and refers to science, education, medicine, arts, religion and agriculture to name a few; he attempted to connect science and religion.

Having originally joined Madame Blavatsky's Theosophical Society, Steiner rose to a prominent position (Collins, 2010, p. 168). In fact, he became head of the German branch of the Theosophical Society in 1902 (Schmidt, 2016), which can arguably be called an international NGO, having its headquarters in Madras, India. Later, he founded the anthroposophical movement in 1912 in Cologne and built the first Goetheanum building. In 1923, he restructured the society through the creation of an Executive Council, becoming its chairman. The Society set up research departments (sections) in the following fields: general anthroposophy; education; medicine; the arts of eurhythmy, speech, drama and music; the literary arts and humanities; mathematics; astronomy; science; the fine arts; social sciences; youth and agriculture (Schmidt, 2016). Anthroposophy is understood as 'a science of the spirit capable of providing us with the ability to comprehend the spiritual as well as the physical nature of life, and to recognise the importance of this knowledge for the future of humanity' (The Anthroposophical Society, 2016a, p. 1).

Steiner attended the Technical College of Vienna and was generally knowledgeable; however, his desire was not the primary advancement of knowledge but the cultivation of knowledge based on individual spiritual knowledge (McDermot, 1996). His knowledge was gained through traditional education and experiments, but he also suggested a supernatural state of consciousness in which humans, including himself, could gain access to knowledge. His students and followers were advised to refer to spiritual practices such as meditations and 'a review', a spiritual ethical thinking. His followers have also experienced unusual state of consciousness in Steiner's presence which was described as a 'revelation'; this might indicate that he had hypnotic abilities (von Keyserlingk, 1999).

Steiner aimed to achieve an understanding of the world in purely philosophical form and concentrated on the study of organic forms of life. He recognised three classes – plants, animals and humans – brought to life by so-called 'etheric formative forces' (antherische bildekrafte). He described how their activity could be studied by the methodical observation of physical phenomena based on Goethe's ideas of plant metamorphosis (Steiner, 1999, 2002). His world view was based on supernatural powers, reflected in his theories about plant development. The notes of his student, K. Waler, show that Steiner taught his students the following:

> The spirits of the elemental world – gnomes, undines, sylphs and salamanders – are actively involved in plant development. They used to be guided and influenced by higher spirits that have now withdrawn from this sphere of activity. . . . The result will be that the spiritual energies of plants decrease, gradually causing them to die off completely. Even artificial fertilizers will not help in this case.
>
> (von Keyserlingk, 1999, p. 8)

Agriculture caught Steiner's attention on two matters – questions relating to nutrition and farming based on anthroposophical ideas. Marie Steiner, his second wife, was an active supporter of his ideas and promoter of eurhythmy – a form of expressive movement arts adopted at Waldorf schools. She recollected his lower Austrian peasant roots and appreciation for the peasant lifestyle (Lindenberg, 2012). It is not surprising that Steiner almost idealised the view of the farm as a living organism, 'a kind of individual entity in itself – a self-contained individuality'; the healthy farm being considered to be self-sufficient in terms of food production (Steiner, 1974, p. 3). It was also a place where people became 'strong . . . simply through the fact that they eat large quantities of their own homemade bread which contains grain from their fields' (Steiner, 1999).

The agricultural problems of today began to appear in the early twentieth century. Farmers in the early 1920s faced the challenge of increasing degeneration in seed stocks and soil. For example, in 1922, German farmers complained to Steiner that some fields produced fewer crops and the food quality of those crops had deteriorated (Pfeiffer, 2011). Ernst Stegemann, a farmer who reported to Dr Steiner these problems, received an explanation that all food cultivars (strain originating and persistent under cultivation) would degenerate over a short period of time at the end of the Kali Yuga period. This required the breeding of new strains, and Steiner gave him directions as to how to breed a new cereal plant from particular grasses in the production of bread, using love, hope and faith and one-grained wheat and oat grass seeds (von Keyserlingk, 1999, p. 11).

Since at least the early 1920s, Steiner was asked how to put his ideas into practical agriculture in order to improve agricultural yields. He began to give practical advice in the use of natural manure without artificial fertilisers, later called the biodynamic agriculture method. The manure mix, known as 500, prepared by Steiner's recipe, was applied for the first time in the garden of Sonnenhof in Arlesheim, Switzerland. Pfeiffer described the process of preparing the first 500 mix which was both a technique trial and reference to Steiner's ideas on cosmic forces:

> Dr Steiner turned back and asked for a bucket of water and showed, then, how the content of the horn should be sprinkled into the water and stirred. My walking stick was the handiest available, and so it was used for the stirring. Dr Steiner stressed above all energetic stirring, the formation of vortices, and the quick reversal sod direction which reveal the vortex-forming action of this energetic stirring.
>
> (Pfeiffer, 2011, p. 9)

Compost was used to reanimate the natural forces of nature, shown by his students as the practicality of his school of spiritual science (Pfeiffer, 2011). In his lectures, Steiner discussed the importance of protein and fats for human nutrition. Plants contain protein which they derive from minerals and pass to animals and humans; in return, they need the carbon dioxide produced by humans (Steiner, 1999). In his agriculture course, Steiner focused on the relationship of earth and soil to the formative forces of the esoteric, astral and ego activity of nature and explained the connection of cosmic creative forces with soil, plants and animals (Steiner, 2003). These ideas were then developed by his students.

It is interesting to imagine how Rudolf Steiner could have perceived the idea of transgenic crops. Countess Johanna von Keyserlingk wrote about a conversation with Steiner:

> Won't it be necessary for new power coming from the world of the spirit to create new plant forms, with the aids of human beings? – Dr Steiner: yes. Or that existing plants are improved, among them also plants now considered weeds? – Dr: yes.
>
> (von Keyserlingk, 1999, p. 11)

Furthering development and promotion of ideas on biodynamic science and agriculture, Steiner's students have continued to work in agriculture and plant science and have institutionalised their knowledge on biodynamic agriculture.

In the summer of 1924 in Koberwitz, near Breslau on the estate of Count Karl von Keyserlingk, Steiner delivered eight lectures known today as the agriculture course (Steiner, 2003). The course was attended by 60 people (Carlgren, 1979). Attendees included E. Pfeiffer, von Keyserlingk and other German aristocratic families and farmers.

The agriculture course was developed by Steiner a year before his death. Due to failing health, he could not promote this new set of ideas himself, as was the case for his ideas on education which became known as the Waldorf School. It was strategic that the agricultural course was institutionalised within the anthroposophical movement. As a result, after accepting an offer from Karl von Keyserlingk to run the course on his estate, Steiner nominated Keyserlingk and Stegemann as 'brand champions', recruited the Agricultural Experimental Circle of farmers, appointed a central clearing house and distributed a confidentiality agreement among course participants during the development stage in order that the ideas would not be spread prematurely (Paull, 2011). This management strategy allowed his students to translate the course notes from German into Dutch, English, French and Italian and reflect upon, summarise and later build up new ideas on agriculture. In 1929, the notes were translated into English and distributed to recipients in the UK, Australia, New Zealand and South Africa (Paull, 2011). Before the beginning of World War II, Pfeiffer moved to the US, where he became an active promoter of biodynamic agriculture and interacted with the leaders of the American organic and environmental movements, including Rodale and Rachel Carson (Paull, 2011, 2013).

The term 'biodynamic' was not invented by Steiner but came from his students. Pfeiffer, who popularised the new practice of manure and compost application and published several books on gardening and farming, formed a definition on biodynamic agriculture: biodynamic agriculture is a method of farming and gardening developed in 1922, based on the advice and instructions of Steiner. The name is a reference to 'working with the energies which create and maintain life' and is a combination of two Greek words: bio (life) and dynamics (energy) (Pfeiffer, 2011, p. 7).

In his last years, Steiner's interest in plants encouraged his followers to become engaged in natural sciences and conduct experiments. Two methods were developed – crystallisation by Pfeiffer and capillary dynamolysis by Lilly Kolisko. The Kolisko method used photography to make visible the invisible etheric forces discovered by Steiner and was considered to be both a scientific and artistic method (Trevelyan, 1978). Pfeiffer offered an analytical method using copper chloride, called crystallisation. The author also used illustrations of different crystal structures in his work (Pfeiffer, 1936). The method has been used by the medical establishment in the diagnosis of cancer (Gruner, 1940). Together, Kolisko and Pfeiffer used chromatography to visualise food quality (Demeter, 2016).

Under Steiner's guidance, his students began what can be called seed banks. Dr Wegman brought seeds of local varieties from Sinai, Pfeiffer collected one-grained seeds from the ancient ruins of Rome. Following Steiner's advice to acclimatise the Chinese sweet potato in Europe, his followers ordered sweet potato samplers from China by post. However, the imported Chinese potato varieties turned to have traits of those imported to China from the West and not original Chinese varieties (von Keyserlingk, 1999, pp. 18–19).

The anthroposophists had their own seed mysteries. Steiner spoke about working with seeds 'meditatively' and gave a meditation to Stegemann, the farmer, who shared them with others (von Keyserlingk, 1999, pp. 18–19). Steiner's followers carried on working with seeds. Many of them, including Pfeiffer, von Keyserlingk and Kolisko conducted experiments on seeds in laboratories. Von Keyserlingk claimed that 12 forms (cultivars) were created from a single seed in Hugo Erbe's establishment, and Immanuel Voegle studied phenotypes of spelt (von Keyserlingk, 1999, pp. 18–19).

Biodynamic knowledge was quickly spread through the network of anthroposophists, and in 1928, there were 66 farms and 148 members of the experimental circle (Schilthuis, 1994).

Organic agriculture today is known for a variety of certification schemes. The biodynamic agricultural movement has also produced a wide network of organisations, which can be classified as NGOs, and different labelling schemes. The Demeter symbol was introduced in 1928, and 'Demeter-Wirtschaftsverbund' was founded in 1932 (Demeter, 2016).

Pfeiffer travelled to the US on several occasions, where he received an offer to manage a laboratory in New York and also acquired and managed a farm there. Later, he even served as adviser to the US Department of agriculture (Schilthuis, 1994). To mark his research, the New York-based Pfeiffer Center, a modern NGO, runs various events and training in biodynamic agriculture (Pfeiffer Center, 2016).

In Switzerland, Hans and Maria Mueller developed and practiced 'organic – biological' farming in the 1950s based on the findings on soil fertility by Hans-Peter Rusch. The Rusch-Mueller method was promoted within a local movement which was formalised in the 1970s with the adoption of the trademark, Bioland, which then became one of the largest certifiers in Germany (Haccius and Lünzer, 2000).

In England, despite the existence of its own movement on organic agriculture promoted by Sir Howard and Lady Balfour, anthroposophy and biodynamic movements also spread. D.N. Dunlop invited Carl Mint, a representative of Count Keyserlingk, to speak to an English audience and settle in Northumberland to run an anthroposophical agricultural foundation. Another separate biodynamic association was founded by Lady McKinnon in 1935; in 1944, the two associations were amalgamated into the biodynamic agricultural association and the *Star & Furrow* journal was established (Schilthuis, 1994, p. 43). The journal continues to exist, and its editors occasionally publish news on GM crops in Europe and express their views on GM crops (Brown, 2013, p. 4).

The network of anthroposophical and biodynamic societies exists in nearly 50 countries (Paull, 2013) and diffuses the original ideas of Steiner and their later interpretations. For example, the manure mix recipes published by the first generation of biodynamic scientists such as Pfeiffer and Kolisko (Pfeiffer, 1936; Kolisko and Kolisko, 1978) were then used and adopted by the next generation of biodynamic educators, such as American biodynamic trainer Michael Maltas

(Pfeiffer and Maltas, 2013) and a German farmer named Karl Ernst Osthaus (2010). Both claimed to use biodynamic methods to solve some modern challenges of agriculture, such as growing resistance to pests and radiation in soil.

Steiner's ideas also allow his followers to develop, patent and gain profit from their ideas. For example, in 1977, Christoph Fischer started the 'Rosenheimer Projekt' which produces a mix of microbiota in ceramic (Effective mikroorganismen) to rehabilitate soil. The project is a business enterprise as well as a community initiative which organises events for farmers in Southern Bavaria, Germany (Fischer, 2014a). Fischer, who is a registered entrepreneur, is also an activist who stands against GM agriculture. He participates in GMO-free Europe movements and invites international activists, such as Vandana Shiva, to talk to farmers within his community (Fischer, 2014b).

An enterprise which was co-founded by Steiner, Ita Wegman and Oskar Schmiedel named Weleda has grown from being a hospital using anthroposophical medicine for sick patients into a global corporation. Dr Ita Wegman, a Dutch doctor, opened the Clinical and Therapeutic Institute in Arlesheim in Switzerland in 1921 which then merged with the Futurum AG founded by Steiner and Schmiedel. Within a year, Weleda's product range included 120 different natural cosmetics and pharmaceutical products. Its name is an adaptation from the Germanic healer and prophet, Veleda, and Rudolf Steiner personally designed the logo which depicts a stylised staff and Aesculapian snake. It is still in use today (Weleda, 2016). After World War II, the company expanded across the anthroposophical web: international subsidiaries were opened in France, Italy, Sweden, New Zealand and Latin America (Argentina and Brazil). Steiner's idea to use mistletoe, a traditional remedy in herbal medicine, to treat cancer, has been realised in Iscador, a licensed drug in Germany (Murphy, 2001).

All this illustrates that Steiner's ideas are widely applied and updated, uniting people across organic agriculture movements, and garner a high profit which serves as an additional incentive for activists.

Florianne Koechlin's ideas on plants and role in the movements against GMOs

Dr Florianne Koechlin is 'a Swiss biologist and chemist', born in 1948, who is well known as a critic of genetic engineering and for her various books and articles (GMO-free Europe, 2012). Her ideas and social activism make an interesting case to discuss the discourse of nature fundamentalism.

She studied chemistry in Middlebury College, US, and biology in the University of Basel. She names plants, stones, water and landscapes as her research interests. While 'Science is [her] world', she is also very artistic and philosophical in her approach, bringing her sketchbook while she walks in nature. She tries to make these inner experiences and images find their own language and to discover answers to almost existential questions ('to see what a fig tree or a mountain stream has to do with me'). She has been using scientific research and painting as inspiration to find alternative knowledge 'about what constitutes life' (Koechlin, 2016).

Alternative science and a new paradigm

In her work, Koechlin makes references to a number of scientific publications and experiments and accompanies these references with her own socio-philosophical interpretations. Her long-term reflection on biotechnology and, particularly, genetic manipulations on plants has raised a number of fundamental issues such as the nature of knowledge and science and what the future agriculture should look like. Paradoxically, she might argue that her work, which I classify under nature fundamentalism, is against science fundamentalism.

She revises the traditional understanding of the gene as a 'section of the DNA, which contains the information for one protein'. Such a definition means a dogmatic view on genes as 'a one-track street from gene to protein to everything else', removing context dependency. She calls it a 'seductive paradigm' which explains gene mechanism with clarity and simplicity. Yet she argues that context matters and there is 'an interactive communication in all directions' when genes talk to proteins, genes and the outer environment. This systematic view on genes, in an arguably biodynamic style, means that the question 'What is a gene?' cannot be answered easily. There are overlapping genes, jumping genes, silent genes, genes with many different functions – the definition of a gene is getting more and more elusive. Genes are part of a dynamic network (Koechlin, 2005a).

This citation can also be used as an illustration of how Koechlin's views on plants and biology generally match the biodynamic point of view. When asked directly in an interview, Koechlin recognises that anthropocentrism is particularly ill reputed in biology, and yet there is something useful to learn, such as the focus on individuality and the role of context (Koechlin, 2006). Her discussion of the underground network that describes the connectivity of plants, soil and microorganisms is very compatible with the biodynamic model (Koechlin, 2015).

This disagreement on the nature of genes (when genes are not considered 'the book of life' but are seen as 'a random collection of words from which a meaningful story of life may be assembled') then becomes the foundation for her critique of gene manipulation, particularly in plants (Koechlin, 2015, p. 3). It also brings the fundamental issue of discussing reductionism in modern biology into the debates on GM crops.

Under the complexity of the system of genes, Koechlin argues that gene transfers are neither precise nor predictable. Scientists operating gene transfers do not know how many genes to integrate into genomes in order not to break 'subtle networks in the cell' and how this could 'change neighbour-relations or cause reciprocal actions'. She refers to pleiotropy as 'the fact that one and the same gene can have different effects'. This makes genetic manipulations in plants particularly complex and possibly dangerous due to the high variety of their genes (Koechlin, 2015, p. 3). She calls the existing literature on pleiotropy 'inadequate' because it does not take into account much of the unintended changes in transgenic plants and poor sampling: 'plants which have poorer "values" are discarded in the laboratory' (Koechlin, 2015, p. 3).

The oversimplified approach to gene function causes Dr Koechlin to doubt the success in creating long-term characteristics in transgenic plants, such as drought and salt tolerance. As the level of stress tolerance in plants is based on the interplay between gene networks, so the focus of genetic engineering on just one component of a complex system is questionable (Koechlin, 2005b).

As a result, the general assessment of modern biological knowledge about the nature of plants is considered to be 'too brief' and like 'dinosaur technology' when scientists do not know in which complex processes they are intervening and which genetic networks can be affected. Modern biogenetics is called 'linear and mechanistic'. Koechlin calls for a different approach which would include 'insights into immensely complex dynamic networks, whose cells, organs and living beings enable them to flexibly adapt to environmental changes' (Koechlin, 2005b).

An approach adopted by Koechlin to understanding genes, which different from classic scientific ones, then leads her to reconsider epigenetics – the study of heritable changes – and evolutionary determinism – a set of views which argue that changes in gene frequencies are by directed or deterministic processes, in contrast with change due to random or stochastic processes (Allaby, 2010, p. 142).

Koechlin re-questions monogenetic hereditary diseases, claiming that only 2% of all illnesses are monogenetic and that the concept of 'genetic illness' itself is becoming a mammoth category, since the causes of all illnesses lie in environmental factors where a gene has played its role. On the one hand, she still agrees that genes are important in causing illness, while on the other, she suggests a reconsideration of what is essential – either it is genes or it is context. Thus, genetic medicine is criticised for ignoring the level of adaptability of epigenetic systems and removing the previous focus on environmental factors and mental well-being (Koechlin, 1999). An overly simplistic approach causes scientists and audiences to forget the feelings of those who are born with such diseases.

In terms of genetic determinism, Koechlin disagrees with the dominant genetic paradigm that genes always stay the same and contain the information for protein and direct the composition of cells and organisms and that information runs only from genes to their environment (Koechlin, 2001).

For her part, she is convinced that information can go in both directions, that genes can also receive information. Therefore, the same gene would have different functions in different environments. To illustrate this, she uses the example that one gene can be responsible for two different traits: causing a visual pigment in fruit flies and beginning the development of the immune system in mice (Koechlin, 2001). Thus, she argues for the refocusing of attention from genes to their interactions with their environment: the central element is that these dynamic networks of genes and proteins have their own beings – they follow network rules that are not denoted within the genes. It requires a new point of view to study these relations. The genes are not in the centre anymore but are cells and organisms, as a whole, including their interactions with different parts. Next she asks if that change in understanding the role of genes might be enough for a paradigm shift in science (Koechlin, 2001).

The main science under question is biogenetics and also synthetic biology which set its sights on creating living beings and is criticised for its standardisation and automation in creating biological systems (Koechlin, 2010). Koechlin refers to three main approaches in synthetic biology: the Chassis-model (the top-down approach), the Lego-model (the bottom-up approach) and the synthetisation of DNA sequences. Synthetic biology is criticised for its pretence to create new life and is opposed in philosophical terms: she asks how 'new' can be created out of nothing.

Life, she argues, is more than just a sum of genes, and it cannot be created. It cannot be created, because it creates itself. She refers to the work of Chilean biologists who defined autopoiesis: Maturana and Verala in the 1970s and early 1980s used the concept of autopoiesis to describe autonomous closed systems, that is, autopoietic. These systems are defined as a unity, 'as a network of processes of production (transformation and destruction) of components that produces the components which: (i) through their interactions and transformations continuously regenerate and realize the network of processes (relations) that produced them' (Maturana and Varela, 1980, p. 79). In Koechlin's view, the Chilean biologists describe the order of life with their concept, since life has the precondition to organise itself, as it is the process of cognition and communication (Koechlin, 2010). This systemic approach allows her to continue to attack biogenetics with its 'mechanist thinking' and 'linear explanation' and offer an alternative way of scientific thinking which she calls 'systems biology' (Koechlin, 2003, 2010).

Along with these complex theories, synthetic biology came along in the 1980s and gave 'the simple gene dogma' a 'triumphal procession'. It has brought back what those who are of the same mind as Koechlin have hoped was gone – the reductionist worldview. It has changed how humans see themselves as Homo creator (Koechlin, 2003, 2010). This then raised the debate on the responsibility of science. Koechlin argues against determinism in modern science: 'What science defines as real, is real in its consequences' (Koechlin, 2010, p. 13).

She distinguishes two major models within the theories explaining life emergence: the mechanic-functional models which follow Descartes's materialism and is exemplified by biogenetics and synthetic biology and the holistic models which 'assume an inexplicable "secret" at the core of every life' and can be translated as 'unresolved emergence' (Koechlin, 2010).

The old 'simple gene dogma' is opposed by the new 'paradigm of life', according to Koechlin. She discusses how a paradigm shift occurs, with reference to six phases of paradigm shift defined by Kuhn. Kuhn described the following phases in this process: a normal science, based on a dominant discourse, faces anomalies which then lead to a scientific crisis; once the crisis is resolved, a new paradigm emerges (Kuhn, 2012).

In its application to molecular biology Koechlin complains that Kuhn's cycle has yet to be realised and depends on 'a decision to take up another paradigm'. At the moment, the call for a new scientific agenda has drowned and the majority of the scientific community has turned back to its former paradigm (Koechlin, 2001). The spread of biogenetics is also a change in paradigm: biology has gained faith again, and in terms of the old debate between heredity versus environment, the pendulum has swung strongly towards genetics (Koechlin, 2001). The new change in paradigm might come as 'the anomalies' are increasing, and the fact that the human genome project found only 30,000 against the projected 100,000 genes is cited as a paradigm crisis (Koechlin, 2001).

It is also interesting to mention that distinguished modern biologists also go beyond the reductionist approach. For example, Professor Marc W. Kirschner, Chair, Department of Systems Biology at Harvard Medical School, explains the complexity to argue that 'an autonomous organism depends on a foundational component of life that never arose before in the physical world – a sense of purpose'. He also accepts that 'a cell talks to a cell' (Kirschner, 2017). But at Q&A session after his lecture, when I ask him about safety of GM crops, he referred to the argument I often received from biologists that genetic modification in plants is similar to selective breeding.

On dignity of plants and how knowledge is created

Dr Koechlin is particularly known for her controversial statements that plants have dignity and that this presupposition should be reflected in legal terms. She has published extensively on this subject, and in January 2016 gave a TED talk about sensitivity and consciousness of plants. She summarised that plants can smell, see, feel touch and hear (Koechlin, 1999). Very much in Steiner's style, she describes that animals and humans have three types of neurobiological cells: sensory cells, nerve cells and brain cells, while plants have at least sensory cells which allow them to receive and respond to signals (Koechlin, 2015).

These communicative qualities in plants are exercised when they are attacked by predators. Plants recognise what kind of predator insects attack them and react by attracting insects that kill these predators or use other techniques. For example, the tobacco plant increases its neurotoxin nicotine level in its leaves when attacked by insects. In the case of tobacco hornworms which are not affected by nicotine, the plant produces another toxin and attracts special ichneumon wasps which prey on hornworms (Koechlin, 2012). With their roots, plants are inbuilt into a wide communication system, the mycorrhiza web, which includes soil microorganisms and other plants (Koechlin, 2013). Plants also appear to recognise each other and almost adapt a social behaviour (Koechlin, 2013, 2014b). This suggests that humans are wrong in seeing plants as isolated and non-communicating objects. Koechlin is ready to step over scientists' fear of 'esoterism' and 'humanize' plants. Under her logic, as plants are not objects, but rather subjects, they may also concede rights. While this does not mean that humans cease eating plants, the consideration of their dignity and rights should be reflected in how they are used in industrial production (Koechlin, 2013, 2014b).

Such theories that plants should have rights sound radical to the mainstream scientific community. Dr Koechlin entered into a polemic debate in a scientific journal to defend her arguments. What is also very interesting is that, while reading Steiner, whose work indeed provides a source of reference to Koechlin's ideas,

I found the following paragraph which suggests that Steiner would disagree: 'Such scientific opinions are not great shakes, they are simply nonsense. Plants do not feel. Nor are plants able to move freely' (Steiner, 1999, p. 250).

With an understanding of the work of the social anthropologist Jeremy Narby, Koechlin has extended her understanding of plants as conscious beings, and she and Narby have started a new collaboration.

Jeremy Narby first travelled to Quirishari in Peru in 1985 as a part of his doctorate fieldwork to study indigenous knowledge of ecology in South America. There he met with local shamans called ayahuasqueros, the drinkers of ayahuasco, and tried to embrace their 'occult science'. He was served ayahuasco by a local ayahuasquero and wrote of his experiences under the effect of the hallucinatory drink. Upon his return to Europe, he continued to analyse how Peruvian shamans receive their knowledge. This led him to discuss another kind of science in which shamans directly receive knowledge from hallucinogenic plants (Narby, 1998, p. 14). In addition, he discussed how the arrogance of the Western scientific community would not be able to accept the knowledge produced by South Americans. He claimed that they do have access to real knowledge and illustrated that with the example of curare, whose efficiency is confirmed by Western scientific empirical research.

In the following years, Narby tried to make sense of the images he saw during his ayahuasco session, and it appears that the traditional images seen in the hallucinations resemble the structure of DNA. A comparison and detailed analysis of ancient symbols across world civilisations have caused him to conclude that this is a global phenomenon, leading him to discuss the existence of universal knowledge on existence, the creation of life and limitations of Western science which appear ignorant and ready 'to belittle the unknown' and 'consider that 97 percent of the DNA in [human] body is junk' (Narby, 1998, p. 38). He called it 'cowboy science' which ignores that 'junk DNA' could have certain functions and is not objective as it claims (Narby, 1998, p. 38). He discussed how Western biology claims knowledge: it 'tends to project its presuppositions onto the reality it observes, claiming that nature itself is devoid of intention' (Narby, 1998, p. 140). With that, he is close to Koechlin, who admits the relativity of scientific statement and offers a space for other experiences, artistic and religious, from intuitive and subjective areas that can create 'non-catchable' knowledge. Koechlin accepts that Peruvian shamans have contacted DNA with a snake-like appearance (Koechlin, 2002a).

Other biologists have expressed an interest in experiencing shamanic knowledge. In 1999, three molecular biologists travelled to Peru to meet ayahuasqueros. One asked the shamans about GM plants, 'if it was appropriate to add genes to plants to make them resistant to disease'. The 'mother of tobacco' answered that 'manipulating tobacco's genome was not a problem in itself, so long as the plant could play its fundamental role in an adequate environment, and so long as it was in keeping with environment' (Narby, 2001, p. 303).

In 2001, Jeremy Narby, with his work 'Shaman and scientists', and Florianne Koechlin, with her presentation entitled 'On agronomic problems based on ecological integrity', participated in an Ifgene workshop called 'The Intrinsic Value and Integrity of Plants in the Context of Genetic Engineering' at Goetheanum, Headquarters of Anthroposophical Society in Dornach (Science Group, 2001).

Associated campaigns

Florianne Koechlin is an interesting example of the anti-GM plants activist. She has been involved in the movement for more than 30 years. She has been involved in politics since 1968 and always in the NGO area. Her main focus has been on genetic engineering, as she came into ideological confrontation with 'agro engineering' (Koechlin, 2014a). In 2014, at the conference Farbe der Forschung in Berlin, it was announced that Dr Koechlin intended to spend more time on her research and detach from the anti-GMO campaigns.

To use the allegory of the mycorrhiza web, Dr Koechlin is well connected to a network of activists, social groups and NGOs opposing GM crops. During her activist career, she was involved in such organisations as an independent ethics committee by the Swiss Federal Council (the Swiss Federal Ethics Committee on Non-Human Biotechnology), EKAH (Ethikkommission für Biotechnologie im Ausser-humanbereich), a number of NGO networks discussing biotechnology, such as the Swiss Working Group on Genetic Technology, known as SAG (Schweiz Arbeitsgruppe Gentechnologie: Dachorganisation aller gentechkritischen schweiz), NGOs GENET and GMO-FREE Europe. She has served as a member of the board to SWISSAID and the Foundation on Future Farming (Stiftungsrat Zukunftsstiftung Landwirtschaft) in Germany (Swisscanto, 2009). She has also worked for WWF Switzerland. Together with her colleagues from SWISSAID and WWF, she organised the international symposium 'Patents, Genes and Butterflies' for the wider public and media on 20–21 October 1994 in Berne, Switzerland. The proceedings were expanded into a book entitled *Life Industry: Biodiversity, People and Profits*. The contributing authors included academics as well as activists such as Patrick Money and Vandana Shiva and made references to such organisations as GRAIN, PAN, and Third World Network to name a few (WWF, 1996). She was in contact with other activists who shared news and their own investigations. For example, in her article on the Mexican maize scandal, she refers to 'journalists' Jonathan Matthews and Andy Rowell who, in addition to journalism, act as a founder of GMWatch, a British anti-GMO NGO, and a member of Greenpeace respectively (Koechlin, 2002b). She has set up her own database of contacts to distribute newsletters updating recipients on the news in the development of GM crops.

At the moment, she presents herself as the director of the Blueridge Institute (Blauen Institute), an NGO located in Muenchenstein, Switzerland. Its purpose is to 'explore new scientific findings about plants and other living beings (particularly communication between plants and their use of networks) and new concepts for agriculture and research strategies for this purpose' and translate 'expert knowledge into concepts that are easily understood by the general public' (Blauen Institute, 2016). In 2016, she delivered a TED talk. However, the most famous campaign that Koechlin has taken part in is the Beobachter Initiative. It is remarkable for several reasons. First, it is one of the earlier examples of European civil society opposition to GM plants. Second, the campaign has shown that civil society groups can exercise a significant influence over the general public, media and policymakers since they have risen as an independent influential actor in public sphere.

The people's initiative 'Against the Abusive Use of Techniques of Reproduction and Genetic Manipulation in Humans' was launched in 1985 by the journal *Der Schweizer Beobachter*. The initiators of the campaign drew attention to the lack of Swiss constitutional regulation on the subject and were inclined to provide respect to human dignity and family protection. By the time the debate had started, the Swiss Academy of Medical Sciences had formulated its own relevant medical and ethical directives (Annee Politique Suisse, 2016).

In 1985, the first Swiss test-tube baby was born and church, conservative and feminist groups started to raise their concerns over self-regulation given to the work of scientists and medics and demanded measures to prevent abuse, calling for the introduction of state control over such processes. This was the beginning of the public debate over the use of biotechnology in Switzerland, and it has become known as the first phase of the Beobachter Initiative. The opposition from these various groups started to form and set the target on Basel, the Swiss region which has a high concentration of pharmaceutical and chemical companies, the main biotechnology users (Bonfadelli et al., 1998, p. 147).

In 1986, a fire in the Sandoz chemical plant in the suburbs of Basel created a new impetus for consolidating the opposition movement to demand better protection for the local population. Members of another civil movement, the anti-nuclear movement, joined the new lobby group against biotechnology. An NGO, Swiss Action Group on Gene Technology (SAG), to which Koechlin belonged, took the lead and opposed the construction of a new biotechnological plant by Ciba-Geigy and the release of GM potatoes in 1991 (Bonfadelli et al., 1998, p. 147).

The Beobachter Initiative is often mentioned in the studies discussing the direct democracy processes in Switzerland. The result of the initiative was the submission of a proposal in 1987 and the Amstad Commission, an expert commission set up in 1986 by the Swiss Federal Council. In 1989, the Federal Council released its counter-proposal to the Beobachter Initiative which accepted the majority of the demands from the Beobachter initiators but suggested that the future constitutional article be broader and more precise. It did not have any prohibitions and incorporated an article on the extra-human domain (Bonfadelli et al., 1998; Année Politique Suisse, 2016). In May 1992, 73.8% of Swiss voters supported the counter-proposal which became a constitutional article. However, the turnout was low (38.6%), and results were similar across Swiss regions.

Despite the approval of the new constitutional article on the use of biotechnology, the public debate became even more polarised. In the same year, SAG launched a new campaign to prohibit any genetic manipulation in humans and other living organisms. The new initiative, known as the Gen-Schutz Initiative (Gene Protection Initiative), brought green politicians and NGOs on one side and academics and the corporate sector on the other. Scientists came to realise that they had to communicate their research to the public if they wanted to defeat the initiative. As a result, they held informal gatherings, used their own already existing structures such as the Swiss Federal Institute of Technology and the Swiss National Science Foundation, and set up their own NGOs such as the Gen Suisse Foundation (Koenig, 1998, p. 1685).

In 1996, the government finally formulated its proposal to accelerate legislation on the use of biotechnology on the extra-human domain. In 1997, genetically modified soya was released into the Swiss market by Monsanto, and the debates on GM in Swiss media reached their peak (Bonfadelli et al., 1998, pp. 148, 153). Both corporate sector and civil society mobilised their resources.

Allegedly, the Swiss pharmaceutical industry spent $10m on the campaign explaining the importance of biotechnology. In parallel, SAG and other NGOs managed to collect 111,000 signatures (Koenig, 1998, p. 1685). Dr Koechlin was approached by media as a leader of the anti-genetic opposition, and media found an interesting fact that her family owned the Geigy Corporation (FT, 1998, p. 9).

In June 1998, Swiss voters rejected the Gen-Schutz Initiative by a large majority (66%). The difference between conducting the Beobachter Initiative and the Gen-Schutz Initiative was the public involvement of scientists. In 1998, around 3,000 scientists and their supporters marched in Zurich, four Swiss Nobel Prize winners held a news conference to oppose the initiative and scientists wrote in the media.

Koechlin explained in the interview the results of the voting by the fact that corporations and scientists explained the use of biotechnology to the general public as 'a medical issue', raising hopes that biotechnology could provide 'hopes for cures' (Olson, 1998). In her later publication, she explained the failure of the campaign as being through the opponents presenting the anti-genetic group as 'die-hard fearful naysayers' who go against 'hope' and the 'future' (Koechlin, 1999). She referred to the famous book *Risk Society: Towards a New Modernity* written by German sociologist Ulrick Beck (1992) which describes modern society as a 'risk society' facing risks with ignorance. She also referred to the American activist Jeremy Rifkin, making a similar argument of the irresponsibility and unjustified authority given to state, science and business in deciding upon the use of genes (Beck, 1992).

Literature

Allaby, M. (2010) *Dictionary of ecology*. Oxford: Oxford University Press.

Année Politique Suisse (2016) Dossier Fortpflanzungsmedizin www.anneepolitique.ch/de/procreationassistee.php as viewed 13.04.2016.

Beck, U. (1992) *Risk society: Towards a new modernity*. London: Sage.

Blauen Institute (2016) About us. www.blauen-institut.ch/s2_blue/pg_blu/pf/a_f.html as viewed 13.04.2016.

Brown, P. (2013) From the Biodynamic Association Chairman. *Star and Furrow*, 120: 4.

Bonfadelli, H., Hieber, P., Leonarz, M., Meier, W.A., Schanne, M., Wessels, H.P. (1998) Switzerland, in J. Durant, M.W. Bauer, G. Gaskell (eds.) *Biotechnology in the public sphere: A European sourcebook*. London: Science Museum, pp. 144–156.

Carlgren, F. (1979) *Rudolf Steiner and anthroposophy*. London: Rudolf Steiner.

Collins, H.T. (2010) *God: The perverted realism, the universal fraud: Part Two*. Pittsburg: RoseDog Books.

Cook, G. (2004) *Genetically modified language: The discourse of arguments for GM crops and food*. New York: Routledge.

Demeter (2016) History. www.demeter.de/what-is-demeter/history as viewed 13.04.2016.

Financial Times (1998) *Swiss hold referendum about genetics: Comment and analysis*. 23 May 1998. L.: 9.

Fischer, C. (2014a) *War ist EM?* Sochtenau: Christoph Fischer GmbH.

Fischer, C. (2014b) Interview with Cristoph Fischer. Berlin, 8 March 2014.

GMO-free Europe (2012) Speakers. www.gmo-free-regions.org/past-gmo-free-conferences/gmo-free-conference-2012/speakers.html as viewed 13.04.2016.

Gruner, O.C. (1940) Experience with the Pfeiffer crystalization method for the diagnosis of cancer. *The Canadian Medical Association Journal*, 43(2): 99–106.

Haccius, M., Lünzer, I. (2000) Organic agriculture in Germany, in S. Graf, H. Willer (eds.) *Organic agriculture in Europe: Results of the internet project*. Bad Dürkheim: Stiftung Ökologie und Landbau, pp. 109–128.

Kass, L. (2002) *Life liberty and defence of dignity: The challenge of bioethics encounter*. San Francisco: Encounter.

Kirschner, M.W. (2017) Clare Hall King lecture. Cambridge, 28 April 2017.

Koechlin, F. (1999) Mythos Gen. www.blauen-institut.ch/s2_blue/pg_blu/pf/a_f.html as viewed 13.04.2016.

Koechlin, F. (2001) Zu wenig Gene, um alles zu erklären. www.blauen-institut.ch/s2_blue/pg_blu/pf/a_f.html as viewed 13.04.2016.

Koechlin, F. (2002a) Ist Naturwissenschaft mit Schamanenwissen vereinbar? Wie Schamanen wissen. www.blauen-institut.ch/s2_blue/pg_blu/pf/a_f.html

Koechlin, F. (2002b) Kontamination durch Gentech-Pollen: Kesseltreiben gegen zwei Forscher. Der mexikanische Mais-Skandal. www.blauen-institut.ch/s2_blue/pg_blu/pf/a_f.html as viewed 13.04.2016.

Koechlin, F. (2003) Dieter Roth, Quantenphysik und 'nicht fangbare Fische'. www.blauen-institut.ch/s2_blue/pg_blu/pf/a_f.html

Koechlin, F. (2005a) Workshop A1: Basics of GM technology. www.blauen-institut.ch/s2_blue/pg_blu/pf/a_f.html as viewed 13.04.2016.

Koechlin, F. (2005b) Fallbeispiel: Transgene dürre- und salztolerante Pflanzen. www.blauen-institut.ch/s2_blue/pg_blu/pf/a_f.html as viewed 13.04.2016.

Koechlin, F. (2006) Die geschundene Kreatur. *Zeit*, 9. February 2006. www.zeit.de/2006/07/N-Pflanzenethik as viewed 13.04.2016.

Koechlin, F. (2010) Vom Homo faber zum Homo creator – oder doch nicht? www.blauen-institut.ch/s2_blue/pg_blu/pf/a_f.html

Koechlin, F. (2012) Die List der Hirse Pflanzen neu denken – Ökologische Impulse für die Agrarkultur. www.blauen-institut.ch/s2_blue/tx_blu/tp/tpf/f_list_der_hirse.pdf as viewed 13.04.2016.

Koechlin, F. (2013) Pflanzenkommunikation. 'Das unterrirdische Internet'. www.blauen-institut.ch/s2_blue/tx_blu/tp/tpf/f_fokus-d.pdf as viewed 13.04.2016.

Koechlin, F. (2014a) Interview with Florianne Koechlin. Berlin, 8th March 2014.

Koechlin, F. (2014b) Pflanzenpalaver und Vetternwirtschaft. www.blauen-institut.ch/s2_blue/tx_blu/tp/tpf/f_interview-pflanzenpalaver.pdf. as viewed 13.04.2016.

Koechlin, F. (2015) Pflanzen haben eine Art iunterrirdisches Gehirn. www.blauen-institut.ch/s2_blue/pg_blu/pf/a_f.html as viewed 13.04.2016.

Koechlin, F. (2016) Willkommen auf meiner Homepage. www.floriannekoechlin.ch/ as viewed 13.04.2016.

Koenig, R. (1998) Voters reject antigenetics initiative. *Science*, 280(12): 1685.

Kolisko, E., Kolisko, L. (1978) *Agriculture of tomorrow*. Bournemouth: Kolisko Archive Publications.

Kuhn, T.S. (2012) *The structure of scientific revolutions*. Chicago: Chicago University Press.

Lindenberg, C. (2012) *Rudolf Steiner: A biography*. Great Barrinton: Steiner Books.

McDermott, R. (1996) *The essential Steiner*. San Francisco: Floris.

McDermott, R. (2015) *Steiner and kindred spirits*. Great Barrington: Steiner Books.

Maturana, H.R., Varela, F.J. (1980) *Autopoiesis and cognition: The realization of living*. Dordrecht: D. Reidel Publishing.

Murphy, C. (2001) *Iscador: Mistletoe and cancer therapy*. New York: Lantern Books.

Narby, J. (1998) *The cosmic serpent, DNA and the origins of knowledge*. London: Victor Gollancz.

Narby, J. (2001) Shamans and scientists, in J. Narby, F. Huxley (eds.) *Shamans through time: 500 years on the path to knowledge*. London: Thames & Hudson, pp. 301–305.

Olson, E. (1998) Swiss reject limits on testing genetically altered animals. *New York Times*, 7 June 1998.

Osthaus, K.E. (2010) *The biodynamic farm*. Edinburgh: Floris.

Paull, J. (2011) The secrets of Koberwitz: The diffusion of Rudolf Steiner's agriculture course and the founding of biodynamic agriculture. *Journal of Social Research and Policy*, 2(1): 19–29.

Paull, J. (2013) The Rachel Carson letters and the making of *Silent Spring*. *SAGE Open*, July–September 2013: 1–12. http://sgo.sagepub.com/content/spsgo/3/3/2158244013494861.full.pdf as viewed 13.04.2016.

Pfeiffer, E. (1936) *The formative forces in crystallisation*. London: Rudolf Steiner Publishing.

Pfeiffer, E. (2011) *Pfeiffer's introduction to biodynamics*. Edinburgh: Floris.

Pfeiffer, E., Maltas, M. (2013) *The biodynamic orchard book*. Edinburgh: Floris.

Pfeiffer Center (2016) About us. http://www.pfeiffercenter.org/about_us/index.aspx as viewed 12.06.2017.

Schilthuis, W. (1994) *Biodynamic agriculture*. Edinburgh: Floris.

Schmidt, R. (2016) History of the anthroposophical society. www.goetheanum.org/Overview.481.0.html?&L=1 as viewed 13.04.2016.

Science Group of the Anthroposophical Society in Great Britain (2001) Newsletter – September 2001. www.sciencegroup.org.uk/sngnl901.htm as viewed 13.04.2016.

Steiner, R. (1974) *Agriculture: A course of eight lectures*. London: Bio-Dynamic Agricultural Association.

Steiner, R. (1999) *From Beetroot to Buddhism answers to questions, sixteen discussions with workers at the Goetheanum in Dornach between 1 March and 25 June 1924*. London: Rudolf Steiner Press.

Steiner, R. (2002) *From sunspots to strawberries: Answers to questions: Fourteen discussions with workers at the Goetheanum in Dornach between 30 June and 24 September 1924*. Forest Row: Sophia Books.

Steiner, R. (2003) *Agriculture: An introductory reader*. Forest Row: Sophia Books.

Swisscanto (2009) Personal details of Florianne Koechlin. www.swisscanto.ch/ch/en/retail/nachhaltigkeit/partner/nachhaltigkeitsbeirat/koechlin.html. as viewed 13.04.2016.

The Anthroposophical Society (2016a) The anthroposophical society. www.anthroposophy.org.uk/pdf/membership.pdf

The Antroposophical Society (2016b) About the antroposophical society. www.anthroposophy. org.uk/pages/about.php as viewed 31.03.2016.

Thesaurus (2016) Fundamentalism. www.dictionary.com/browse/fundamentalism?s=t as viewed 13.04.2016.

Trevelyan, G. (1978) Foreword to the second edition, in E. Kolisko, L. Kolisko (eds.) *Agriculture of tomorrow*. Bournemouth: Kolisko Archive Publications, p. 1.

von Keyserlingk, A. (1999) *Developing biodynamic agriculture: Reflections on early research*. London: Temple Lodge.

Weleda (2016) Since 1921 Weleda: A company inspired by anthroposophy. www.weleda. co.uk/our-heritage/since-1921/page/since-1921 as viewed 13.04.2016.

WWF (1996) *Life industry: Biodiversity, people and profits*. Gland: WWF.

3 The colonial discourse

My Indian friend in Cambridge threw us a wonderful dinner, consisting of pop-padum with chutneys, chicken curry, and a pudding which was Eton mess and she jokingly served it to us saying 'I should not forget my colonial past'.

Introducing the discourse

The initial inspiration for this chapter has come from the title of the article 'India After Gandhi – From the British Raj to Monsanto Raj' by Neha Saigal (2013).

The colonial and post-colonial discourses are connected to one another and bring up a number of relevant themes to the discussion of GM crops, such as differences in perception, implementations and consequences of GM crops in developing and developed countries, management of common pool resources, innovations and transfer of knowledge in science and agriculture, food sovereignty and Western and indigenous knowledge.

A definition of colonialism is as follows:

> A process by which European nations found routes to Asian, African, and South American regions; conquered them; undertook trade relations with some of the countries and kingdoms; settled for a few centuries in these places; developed administrative, political, and social institutions; exploited the resources of these regions; and dominated the subject races.
>
> (Nayar, 2012, p. 2)

The geography of colonial and post-colonial discourse is two-dimensional: a metropolis and a colony. Most of the themes are then constructed upon the inter-actions of these two geographic subjects. They are based upon the comparison and often opposition of the two – a centre and a periphery.

The geographic location then becomes a locality in terms of hierarchy and dominance. Colonialism is

> characterized by military conquest; economic exploitation; the imposition of Western education, languages, introducing Christianity, forms of law and order; the development of infrastructure for a more efficient administration

of the Empire – railways, roadways, telegraphy; and the documentation of the subject races' cultures (history, ethnography, archaeology, the census).

(Nayar, 2012, p. 3)

Most colonies, particularly India and Africa, discussed in this chapter belong to tropical areas, which results in differences from Europe in terms of the natural environment and way of life. These differences are exemplified in agriculture.

Mbembe, reflecting on post-colonialism, wrote about 'the uncompromising nature of the Western self and its active negation of anything not itself' (Mbembe, 2001, p. 12). Within colonial discourse, colonies represented otherness, allowing Western metropolises to express a 'desperate desire to assert its difference from the rest of the world' and in that instance, Africa has become a metaphor of 'absolute otherness' (Mbembe, 2001, p. 2).

On the one hand, the colonial discourse recognises and celebrates that difference in economic, social and cultural terms. On the other, there is an impulse towards levelling difference, introducing the same standards and rules to colonies as those adapted by the metropolis. Arguably, Mbembe's 'twin project of emancipation and assimilation' used in relation to the African context (Mbembe, 2001, p. 12) can be used as a general description of colonial and post-colonial discourses.

The discourse analysis distinguishes in this case a timeline – colonial rule and the independence era. Paradoxically, instead of the independence period which, through the separation from the metropolis, should have ushered in a new beginning and set of improvements, 'the future horizon is apparently closed, while the horizon of the past has apparently receded' (Mbembe, 2001, p. 16). In comparisons of pre-colonial, colonial and post-colonial states of economic, social and cultural affairs, the pre-colonial period is usually lamented and longed for, the colonial promotes a sense of disgust, while the post-colonial is often taken as a possibility for change, improvement and the severance of the dependency that defined the colonial period. However, a lot of themes in post-colonial discourse bring up colonial references, and the total separation has seemed not to have taken place.

Given the fact that 90% of the world's rural population has lived and continues to live in Africa and Asia (Thirtle et al., 2005), and agriculture remains their main occupation source for food subsistence, colonial interference and post-colonial influence on agriculture have meant serious implications for lifestyle and quality of life for indigenous populations living in the colonies.

The inclusion of colonies into one coordinated system of colonial agricultural production and trade resulted in the introduction and spread of a variety of crops across the colonies and metropolises. Tropical plants have provided an abundant cultivar variety. The European intrusion also changed how agricultural resources were managed in the conquered territories. The population in these territories has been engaged in agriculture for centuries and has its own traditional methods. The main motivation for colonisers to intervene in the agriculture of its colonies was to contribute to the imperial wealth of the metropolis and 'the moral justification of the planter's existence'. According to a British agricultural officer in Africa, 'he brought skill and capital and created wealth in areas where it would have otherwise been non-existent' (Masefield, 1950, p. 65).

Agricultural management of colonies reveals such concepts as rationality, modernity and civilisation in which all are delivered by colonisers to the colonised population. Technological innovation in agriculture illustrates all three concepts. For example, in 1953, the Society of Coston University held a symposium to discuss the challenges of colonial agricultural development. Also of concern was the slow progress and profit concerns in the colonised country's agriculture, along with the input of metropolitan scientists (Wallace and Martin, 1954).

When the colonisers left, the issues in agriculture remained, so the major debate in former colonies centred on possible models of post-colonial agricultural development. Two options existed: the first was to return to traditional indigenous agriculture and deny Western knowledge and technology, while the second was to upgrade the economic development to a similar level of the former metropolis and use the same technologies as Western countries. Both models aimed at the same target – independence from the former metropolis, yet they are different in their choice of strategies. And this is the point at which the debate on the use of genetically modified crops in developing countries enters. Under the first model, GM crops are seen as a means of post-colonial dominance and are totally rejected, while under the second, transgenic crops become a strategy for catching up on agricultural development and the possibility to exert independence. India and Africa serve as examples of how the two different models are discussed.

Ideal village: a social agricultural utopia in the colonial discourse

It makes sense to quickly review ideas on post-colonial development and agriculture of the two influential leaders of the early post-colonial period in India and Africa – Mohandas Karamchand Gandhi and Julius Nyerere. These ideas, as will be shown, are both utopian, still very much in use and referred to by representatives of social movements in these two regions and beyond. In this period, Marxism and socialism were almost unavoidable, and discussion on post-colonial development has embraced that. Both leaders referred to in this sub-chapter have made a socialist village a centre of their models of development.

India: Gandhi's views

Mohandas Karamchand Gandhi (1869–1948) formed his model of national development from the harsh critique of India's colonial past. When Gandhi first read R. C. Dutt's economic history of India, which described how thriving crafts in the villages were destroyed under the rule of the East India Company, he wept (Gandhi, 1997, p. 107). He observed and lamented the extreme exploitation of farmers by the foreign government and their own countrymen: '[The villagers] produce the food and go hungry. They produce milk and their children have to go without it' (Gandhi, 1997, p. 140).

Such unfairness and exploitation was caused by the colonial regime and 'Western civilisation', which was based on 'machinery' and considered to be different from India's 'true civilisation' in which the country could evolve (Gandhi, 1997, pp. 66–67). Gandhi's vision was based on the idea of swaraj, first introduced in a pamphlet, meaning control over mind – a spiritual concept of self-discipline and abstinence. In terms of economic development, swaraj was applied as a means to moderate social inequalities, a 'life-corroding competition'. This idea was inspired by Western socialist thinkers such as Ruskin (Gandhi, 1997, p. 68).

Under the influence of Western thinkers such as Ruskin, Tolstoy and Bondaref, Gandhi arrived at the promotion of manual labour as part of the social revolution (Gandhi, 1997, p. 69). By contrast, machinery, which he understood very broadly and included technology such as electricity, telegraphs and telephones, was presented as 'a great sin' and 'a chief symbol of modern civilisation' (Gandhi, 1997, pp. 107, 110, 130). He advocated for a boycott of 'all machines that made goods' (Gandhi, 1997, p. 107). He used the example of the cotton industry to illustrate a possible withdrawal from the use of goods produced with machinery and advocated for hand-woven techniques that had existed for centuries, long before the industrial revolution (Gandhi, 1997, p. 109). Science without humanity was considered one of the seven social sins (Gandhi, 1995, p. 89).

The urbanisation process and decline of rural activities were perceived as tragic, a time to be retrieved. Economic and social prosperity were understood in terms of food sufficiency; food crops were given priority over cash crops (Gandhi, 1997, p. 53). The village was seen as the nucleus of a new model of development based on the ethics of morality and fairness. It was considered to be the economic centre, producing food and goods, but also a political unit providing education and political engagement (seven lakh of villages would make 'a well-living republic' (Gandhi, 1997, p. 63).

The desire of Gandhi to go back to the roots of the Indian lifestyle, including the use of 'the same kind of plough as existed thousands of years ago', and his condemnation of technology can be seen as that of 'a latter day Luddite' (Pinto, 1998, p. 35). The major reasoning for suggesting the return to a pre-colonial style of agriculture were his principles of non-violence and simplicity and the ability of the farmer to control his life: 'The sum and substance of what I want to say is that the individual person should have control over the things that are necessary for the sustenance of life' (Gandhi, 1921, cited in Pinto, p. 17). As in a good traditional socialist worldview, Gandhi's understanding was that peasants and workers should keep control over the means of production of essentials for living, including seeds and land.

To maintain non-violence and a fair share for all, discipline and abstinence in life was promoted. In that vein, one of Gandhi's wisdoms was the moderate consumption of food: 'Eat to live for service of fellow-men. Do not live for indulging yourselves. Hence, your food must be just enough to keep your mind and body in good order' (Gandhi, 1995, p. 24).

He practiced a vegetarian diet and was convinced that the 'vast vegetable kingdom' should be able to provide enough nutrients and liberate people from the

ethical concerns of producing and eating meat (Gandhi, 1995, p. 62). He practiced 'dietetics', or 'experiments in treating diseases with natural curative agents
only such as earth and water and without recourse to drugs' (Gandhi, 2016). This
may sound similar to the view of nature fundamentalists described in Chapter 2,
particularly with ideals relating to the organic and cooperative Vedic society based
on ideas from Varnas, the ancient Hindu literature. Yet Gandhi's ideas are not
nature-centric, depicting an anthropocentric model with a low impact on nature,
although the main focus is given to local values and an economy which is discon-
nected from international entanglement. Environmental concerns are important
but not central to Gandhi's spiritualism and, in certain cases, might be overridden
(Peritore, 1999, p. 61).

Gandhi travelled within the Commonwealth states, analysing and comparing the economic and political development of Britain and its former colonies
of India and Africa. He aspired that India would serve as an example for other
exploited colonies and could lead the way in the post-colonial period by providing spiritual wisdom to developed and developing countries (Gandhi, 1995,
p. 38). Three elements of Gandhiji's philosophy have retained their influence
today – swaraj (home rule), satyagraha (non-violent resistance) and swadeshi
(local self-sufficiency).

Africa: Nyerere and the ujamaa concept

In parallel to Gandhi, Africa produced its own intellectual reflection upon the
models of development suitable for the region. For example, Julius Nyerere,
president of Tanzania from 1961 to 1985, developed a policy and intellectual
framework on socialism and self-reliance. Its main concept was ujamaa which
shares many similarities with Gandhism. Of course, Nyerere is not the only one
to write about the independence and economic development of Africa. In South
Africa, the colonial and post-colonial discourse is based on Nelson Mandela's
aspirations for a fair society (Thirtle et al., 2005). I explore the ujamaa concept, as it summarises the main points of the African post-colonial development
discourse.

Ujamaa is not just an equivalent of the word 'socialism' in Swahili. It was
chosen to emphasise 'the Africa-ness of the policies' and the idea of familyhood
('mutual involvement in the family', Nyerere, 1968, p. 2). Familyhood in this
instance is understood as a commune:

> The traditional African family lived according to the basic principles of uja-
> maa. Its members did this unconsciously, and without any conception of what
> they were doing in political terms. They lived together and worked together
> because that was how they understood life, and how they reinforced each
> other against the difficulties they had to contend with – the uncertainties
> of weather and sickness, the depredations of wild animals (and sometimes
> human enemies), and the cycle of life and death.
>
> (Nyerere, 1968, p. 337)

Many farmers supported the initiative, which they saw as a better alternative to the exploitation of the colonial period. According to a farmer in the Litowa ujamaa:

> The first few months were spent on evening discussions and exchanging ideas around the fire about various aspects of experiments achieved one had while working in the sisal estates, on mission farms or any other experience. . . . Together with the discussions, were highly touchingly political songs. . . . These songs recounted the evils inflicted upon Africans by colonialism. The thought the Uhuru was nearing when everybody was still ignorant and so poor made them weep.
>
> (von Freyhold, 1979, p. 73)

Ujamaa is comparable with Gandhism from its intellectual response to the challenges of the late colonial and post-colonial periods. Like Gandhi, Nyerere wanted to build upon the past and traditions of his country ('to grow, as a society, out of our own roots') while emphasising 'certain characteristics of our traditional organization' he proposed to 'embrace the possibilities of modern technology' (von Freyhold, 1979, p. 73). In his vision, the ujamaa society was to serve man's equality and everyone was expected to work, either in the traditional farm called shamba or in other household activities; working meant to contribute to the total output of goods and welfare through hard labour. People were to be organised in cooperatives and jointly own tools of production (von Freyhold, 1979, pp. 6–7). Agriculture was understood as the basis of development, a way to improve living standards for African people (von Freyhold, 1979, p. 104), and as the base for the socialist regime (von Freyhold, 1979, p. 346). The improvement in living standards was to come from the increase in crop production. Farmers were expected to carry on their hard labour and other members of society, who had appropriate college education, were expected to educate farmers and teach them how to further increase the production from the land (von Freyhold, 1979, p. 105). So the traditional model of ujamaa was based on respect, sharing and obligation to work and the educational part included learning and taking advantage of Western technologies, particularly in village communities (pp. 340, 354). The lack of financial resources available on the continent, however, was recognised as a limitation for technological advance: 'There is no substitute for this [hard work by people], especially as we do not have large accumulations of capital which can be invested in agricultural labour-saving devices or in increased productivity' (von Freyhold, 1979, p. 346).

The ujamaa development was an ambitious project, which was difficult to realise in practice and remained an experiment and social utopia. In 1973–1974, the widespread drought and world oil crisis led the Tanzanian government to drain foreign currency reserves on food imports, food self-reliance was promoted through the vijiji campaigns where the population was forced to grow food crops and villagers had to resettle in larger farming cooperatives (Boesen et al., 1977, p. 170). Without adequate institutional support – there was poor local leadership and sabotage – and bad harvest, several hundreds of Tanzanian ujamaa coopera-tives collapsed. Some continued to exist until their major tools of production were confiscated by the government (von Freyhold, 1979, pp. 72–74).

Modern post-colonial discourse discussing GM crops

Post-colonial discourse is an interesting mix of ideas, all referring to overcoming challenges inherited from the colonial period. However, there is unanimity on how to reach it. It is also the case that while most themes in post-colonial coun-tries are about independence, this is still to be achieved since informal political dominance continues to exist. There is also what can be called a subconscious pat-tern. Despite the target to become independent from the metropolis, intellectuals in former colonies constantly refer to the former metropolis for justification and look up to it as a good example. In this way, colonialism has formally ended and yet survived in the minds of representatives of this discourse.

This part of the chapter will discuss the fact that despite a difference in a major point regarding the use of GM crops, its opponents and advocates operat-ing within the colonial discourse share some common ideas and the difference in their views can be traced back to their original intellectual background, be that Gandhism or the African model of development supporting the introduction of technological innovations. A comparison of the life story and ideas of Calestous Juma and Vandana Shiva, a comparison which they would both probably find very uncomfortable, actually shows that they have much in common.

India: Vandana Shiva's upgrading Gandhi's views to fight against GM crops

Vandana Shiva is often referred as 'the deserving heir to Mahatma Gandhi's leg-acy': 'What Gandhi was to the British Empire, Dr Shiva is to Monsanto' (Popham, 2014). She is one of the most active representatives of the Indian environmental movement who reclaims Gandhi's ideas. She is also one of the most known activ-ists against GMO. In that role, she holds celebrity status:

> She has been called the Gandhi of grain and compared to Mother Teresa. If she personally accepted all the awards, degrees, and honors offered to her, she would have time for little else. In 1993, Shiva received the Right Livelihood Award, often called the alternative Nobel Prize, for her activism on behalf of ecology and women.
>
> (Specter, 2014)

Vandana Shiva was raised in Dehradun, Northern India, the nearest large city to the Chipko movement (Pearce, 1991). She studied physics and earned her doctor-ate in the philosophy of science. After a period teaching in the university system, she succeeded in a career in international activism (London, 2016). Her entry into the Indian environmental movement began with the Chipko movement, whose aim was to protect Indian forests from logging and ensure villagers had access to the natural resources it provided. It practiced the Gandhi methods of satyagraha, or peaceful walks in protest. It included men and women, among those women two Britons who came to follow Gandhi's ideology in India from Britain: Mirabehn (Madeleine Slade) and Saralabehn (Catherine Helman) who played a prominent role. With the movement, Shiva began her activism and developed her views on environmentalism, corporations and national development.

In her publications, she refers to the views of fellow activists, including Mirabehn, and on numerous occasions to Gandhiji's papers (Shiva, 1991). She refers also to Dutti's work, describing the damage caused by the East India Company to the country's economy, as did Gandhi (Shiva, 1991, p. 47). Colonial policy is blamed for the destruction of the original state of Indian agriculture and food security as well as world recession and food crises (Shiva, 1991, p. 27). She cites British colonial administrators such as the British Chief Commissioner to show how colonial rulers understood little about the Indian agricultural model and how the 'cultivating the savages' theme practiced by the British was detrimental to Indian food security. Farmers were denying growing food crops and were forced to move into cash crops needed for the empire, such as indigo and opium (Shiva, 1993, 1991).

With confidence, it can be argued that Dr Shiva follows the socialist views of Gandhi. Where Gandhi set the control of means of production central to his model of Indian development, so does Shiva. Seeds and land are major means of production and independence for farmers. The colonial trend to involve farmers in the market and deprive them of their own means of production – first of all, land – has led to impoverishment and starvation. The pre-colonial system of land ownership and management called bhaicharia allowed the scattering of land and promoted cooperation and collective action (Shiva, 2005, p. 19). In contrast, colonialism disposed peasants throughout the world of their land entitlement. The British introduced the zamindari or landlord system to divert land from growing food to growing other crops in order to extract revenue from its cultivators (Shiva, 1991, p. 47).

From such colonial policies practiced over centuries, 'the poverty of the third world' has resulted in a drain on resources. In modern times, through globalisation and its major actors – corporations – the same methods of depleting sustenance agriculture have been used and even accelerated (Shiva, 2005, p. 17). Shiva is opposed to the market-based economic model. She is not against markets as a place to exchange goods but opposes the 'society [which is] replaced by capital' and directed by the invisible hand of the market and led by 'the anonymous face of corporations'. Under such a model, she argues, farmers are excluded in their role of producers and the needs of the public are not satisfied, as they are overruled by greed, profit and consumerism (Shiva, 2005, p. 18). While under colonialism, land was the main means of production of which farmers were deprived, in the post-colonial period, which could be described as a neo-colonial period, it is seed and biodiversity. She presents her own version of tragedy of commons in which the whole planet becomes a commons where corporations treat natural resources as 'a global supermarket, where goods and services are produced with high ecological, social, and economic costs and sold for abysmally low prices' (Shiva, 2005, p. 2).

That ownership of the rich is based upon the dispossession of the poor through privatising common public good (Shiva, 2005, p. 2) is not new, but in the modern version, it includes genetics – international corporations patent transgenic seed which farmers then need to buy. This means that corporations treat everything as property and life forms are given value on that scale only; it also allows the corporate sector to easily violate farmers' rights: to make them pay for the usage of new seeds (Shiva, 2005, p. 3). This she calls the seed wars waged by agro-business against farmers in India and Africa (Shiva, 2012).

In Shiva's analysis, there is a confrontation between two economic models: one which she calls a 'living economy' against the 'economy of dominance'. A 'living economy' is referred to as a local economy, which is based on the Gandhian principle that 'the earth provides enough resources for everyone's need, but not for some people's greed'. This model is also based on 'earth democracy' – a set of nine principles that ensure equal access for all members of human society to natural resources. The second model is a market-based economy, which 'myopically focuses on the working of the market', ignoring nature's capacity to provide resources for economic growth and not acknowledging the needs of majority of population (Shiva, 2005, p. 10). International institutions that promote a market economy are equal to former colonisers and are considered to be responsible for widening inequalities between the North (metroplises) and the South (colonies) (Shiva, 2005, p. 14). The green revolution is used as an example of the Western paternalistic interference enforcing a model of agriculture on the South with detrimental environmental damage and the further impoverishment of farmers (Shiva, 1991).

Modern agriculture that is widely based on growing monocrops is also considered to be negative, as it removes farmers' self-sufficiency in food production, leaving them vulnerable to speculation from outside. Monocropping also leads to the loss of local cultivars and general biodiversity. Shiva refers to these processes as bio-piracy and bio-apartheid. Following her past engagement in the Chipko movement, Shiva includes forestry as being part of agricultural management which operates on indigenous knowledge: local tribes know how to use different plants which Westerners often consider to be weeds (Shiva, 1993, p. 26). She reminds the reader how fast British oak forests were destroyed in order to meet the demand of the British Empire to build up its fleet (Shiva, 1993, p. 17).

Shiva makes reference to Sir Albert Howard (Shiva, 1991, p. 25), the founding father of British organic agriculture. While Shiva supports the organic model of agriculture and comes very close to nature fundamentalism with the argument on the important 'symbiotic relationship between soil, water, farm and plants' (Shiva, 1993, p. 27), the main focus here is on the original supremacy of indigenous Indian agricultural knowledge. Sir Howard worked as an agricultural investigator in India for 25 years, where he studied traditional composting processes, later promoting these in the UK as organic farming (Howard, 1953). The fact that the British botanist used the indigenous agricultural knowledge of India counteracts the argument of the British Empire's knowledge supremacy.

In her writing, Shiva relies upon Gandhi's stand against modern civilisation which is self-destructing. The indigenous way of life (local cultures), describing how India used to exist in the pre-colonial past, is considered as a strategy to re-establish a balance with nature (Shiva, 2005, p. 51). Also very much in the Gandhian style, Shiva is against industrialisation which brings capital-intense machines and serves as a measure of national development (Shiva, 2005, p. 5). Instead, development based on mutual cooperation and mutual aid, as outlined by the Russian socialist prince Kropotkin, is referred to and promoted within her work (Shiva, 2005, p. 44).

Dr Shiva studied philosophy of science in the US, yet her perception of science can be called very anti-Western. She understands science as 'a product of social forces . . . [in] a privileged epistemological position of being socially and politically neutral', which offers technological fixes for problems, problems that are often created by science (Shiva, 1991, p. 21). Modern Western science is characterised as 'fragmented into narrow disciplines and reductionist categories . . . [with] a blind spot with respect to relational properties and relational impacts' (Shiva, 1991, p. 22). She calls this process the decontextualisation of science where 'the negative and destructive impacts of science on nature and society are externalised and rendered invisible' (Shiva, 1991, p. 22). This then leads to social conflicts, the unethical behaviour of corporations and disappearance of local indigenous knowledge.

She distinguishes two phases in this process of sweeping away local knowledge. First, the knowledge is ignored and its existence denied by dominant Western knowledge which is then promoted as universal. However, it is argued that Western knowledge is also just a local system 'with its social basis in a particular culture, class and gender'. It is 'merely the globalised version of a very local and parochial tradition' emerging from a colonising culture. Thus, she argues, modern knowledge systems are 'themselves colonising' (Shiva, 1993, p. 9). Knowledge system dominance, which is reductionist in its nature, supports the power nexus and capitalism based on its colonial past (Shiva, 1991). GM seeds are perceived as a capitalist knowledge, which offers a false solution to the global problem of climate change and food insecurity (Shiva, 2011). She predicts that GM seeds will yield worse results than the outcomes of the Green Revolution which in her opinion has failed India (Shiva, 2011).

In her activism, Shiva has embraced Gandhian ideas of peaceful resistance and non-violence. From hugging trees under Chipko activities, Shiva moved into international activism and lobbying and suing corporations. Her methods of resistance are Gandhian:

> For us, not cooperating in the monopoly regimes of intellectual property rights and patents and biodiversity – saying "no" to patents on life, and developing intellectual ideas of resistance – is very much a continuation of Gandhian satyagraha. It is, for, me keeping life free in its diversity. That is the satyagraha for the next millennium. It is what the ecology movement must engage in, not just in India, but in the United States as well. People who believe in the freedom of ideas must engage in this wherever they are.
>
> (London, 2016)

Her resistance seeks to prevent the patterns of colonisation by the Western system, as she described it in reference to knowledge systems: first, challenging an issue on moral and ethical grounds, then operating with legal instruments, referencing the Convention on Biological Diversity (Centre for Research on Globalization, 2016). She supported Percy Schmeiser, the Canadian farmer whose fields were 'invaded' by Monsanto's Roundup Ready Canola GM seeds (Shiva, 2000). She was opposed to American corporations, such as W.R. Grace, patenting traditional Indian varieties like the neem tree, a natural pesticide, and basmati rice which had not been patented and were available through open access in India (Shiva, 2001).

Dr Shiva founded several NGOs and participated in international networks of NGOs: The Research Foundation for Science, Technology and Ecology (1982), then Navdanya (1991) which she calls 'a national movement to protect the diversity and integrity of living resources, especially native seed – the promotion of organic farming and fair trade' (The Chopra Foundation, 2016). In 2000, Satish Kumar, another Indian activist who resides in Britain and edits an ecological magazine entitled *Resurgence & Ecologist*, suggested to Dr Shiva that she open an organic farm as a platform to attract Western organic activists and promote organic agriculture in India (Bhatt, 2015). As a result, Dr Shiva also started Bija Vidyapeeth, 'an international college for sustainable living in Doon Valley in collaboration with Schumacher College, UK' (Shiva, 2016, para. 2).

On the premises of a bankrupted farm in her natal Dehradun area lies this organic farm, which now serves as a model farm with solar batteries and a mango orchard blossoming in soil formerly exhausted by eucalyptus. In March 2015, I travelled to India to attend a workshop on pesticides on the Navdanya farm. The initial advert on the website promoted that the course would be led by Dr Shiva and I registered with a manager of the farm. To my surprise, when I arrived, there had been a radical change in plans: there was no workshop led by Vandana Shiva; instead, it was a closed event held by mainly German lawyers from Brod für die Walt (a German NGO: 'Bread for the World') and their Indian and European associates to discuss strategies to win over corporations in legal cases about environmental damage. I was not the only one left in confusion. Two Americans had travelled to attend the workshop and meet Dr Shiva. So we had to be satisfied with a general visit to the farm and attend a couple of general talks given by our new German friends. The tour was nice; in the middle of the farm there is a Rudrakash tree which, according to Indian mythology, is a sacred tree that emerged from Lord Shiva's. The tree has a plaque attached which reads 'Rudrakash planted by His Royal Highness, Prince Charles, Prince of Wales 7th November 2013'. I found this to be a particularly interesting fact in terms of post-colonial discourse.

Calestous Juma's African development model

Although Calestous Juma does not cite Nyerere in his work, his ideas on African development are close to those elaborated by the Tanzanian leader. Anti-GMO activists have presented Prof. Juma as 'a Monsanto friend' and modern coloniser

(Kranz, 2015; Jalees, 2014), yet even they (Jalees, 2014) have noted his critique of colonial agricultural development in Africa. This review of his views will show that many points in the post-colonial discourse of Prof. Juma and opponents to GM crops from developing countries, such as Dr Shiva, share similar views on colonial discourse. In fact, Prof. Juma and Dr Shiva share quite a few things in common, such as postgraduate degrees from a foreign university and involvement in social activism through establishing NGOs.

Prof. Calestous Juma was born and raised in a Kenyan village close to Lake Victoria, on the border with Uganda. He comes from a farming family with a moderate income that sought opportunities to improve their farming and household income. For example, they were eager to grow cassava, which was not an indigenous crop for Kenya but has since become a staple crop. He began his career as a teacher in science in Mombasa and worked as a freelance journalist for a Kenyan environmental group, launching a magazine. Then, he studied for a postgraduate degree in science and technology policy in the University of Sussex (UK) (Allen, 2013).

He has worked as an executive secretary of the United Nations Convention on Biological Diversity and as a chancellor of the University of Guyana. He is currently a professor of international development in Harvard Kennedy School (Juma 2012). In 1988, he founded the African Centre for Technology Studies, an NGO, in Nairobi. He is also connected with the Bill and Melinda Gates Foundation through directing a project funded by the Foundation, and he heads up a research centre in Harvard Kennedy School.

Professor Juma is a GM optimist and advocate for the technological innovations of which the African continent is currently deprived. His general point is that Africa needs technologies to overcome its underdevelopment. This state of development is the result of previous colonial policies. For example, he links the lack of national and local institutional capacities in his home country, Kenya, with the country's colonial past (Juma, 1989, p. 179).

To say the least, colonial policies in Africa were unfair towards the population. The intervention of colonial agricultural management through the introduction of new crops, including tea and maize, changed international trade patterns to favour the interests of the British Empire. In this, it also changed the 'socio-economic fabric' in colonies such as Kenya. Local cultivars were mainly ignored, says Juma (Juma, 1989, p. 180). The post-colonial Kenyan administration continued the same policy of introducing exotic species until after the mid-1980s when 'public perception' began to move towards more engagement with local communities to use and improve local varieties (Juma, 1989, pp. 184–185). The absence of infrastructure, administrative and financial resources has been blamed for poor performance in conserving the local genetic variety (Juma, 1989, p. 188). The colonial neglect of local varieties and confiscation of land and cattle, for example, from Maasai tribes, and the fixing of land boundaries to promote colonial agricultural settlements and discourage African pastoralism greatly affected Kenyan agriculture (Juma, 1989, pp. 186–187). Juma also notes that these post-colonial international policies aiming to support the breeding of local animal and plant varieties have also coincided with a growing interest of developed countries in accessing the genetic resources of developing nations (Juma, 1989, p. 187).

Interestingly, similarly to Dr Shiva, Prof. Juma supports crop diversity in agriculture and sees it as an expression of support of the traditional lifestyle (Juma, 1989, p. 190). He also refers, as does Vandana Shiva, to the fact of soil depletion resulting from colonial practices, although he also acknowledges the first colonial attempts to conserve soil (Juma, 1989, p. 193).

He is not supportive of the idea of complete reliance on the market mechanism in Africa as promoted by international agencies such as World Bank. He argues that innovations in agricultural sector are too nascent to be subjected to the strong selection pressures of supply and demand (Juma, 1989, p. 191). He is aware of the corporate interest of developed countries to use the genetic resources of developing countries. He uses the example of the aloe plant which was researched by the Jodrell Laboratory of Kew Garden (UK) and it is likely that in this collaborative project Kenyan researchers did not receive the full picture of the plant's commercial value (Juma, 1989, p. 198). In 1989, he warned about the absence of Kenyan legislation on commercially useful species which could endanger 'the future availability of such plants', resulting from overharvesting and depletion of wild plants' populations (Juma, 1989, pp. 198–199).

Prof. Juma and Dr Shiva agree on the point about the negative consequences of the colonial regime for the state of development in Asia and Africa. Both have recognised the value of local ecology and traditional knowledge. The point of differentiation starts with the explanation of the current state of underdevelopment in former British colonies. Vandana Shiva blames the market, whereas Juma does not oppose a market economy; it is the lack of venture capital and education that he considers the main obstacle to technological innovations which could fix African development (Juma, 1989, p. 216). In the discussion on possible strategies for socio-economic development for developing countries, Vandana Shiva comes very close to Gandhian ideas of returning to traditional methods in agriculture and self-sufficiency, while Juma, in parallel with Nyerere's argument, suggests that Africa should catch up with Western powers, its former colonisers, by adapting and developing its own technological advances.

Juma refers to the post-colonial period as 'a period of major agricultural and industrial discontinuities requiring national and regional policy changes' and also technological innovations (Juma, 1989, p. 208). For him, biotechnology is considered a source for such a technological frog leap, a revolution. He is also aware of the challenges facing the Green Revolution but thinks that the new biotech revolution will be beneficial for Africa. Biotechnology, he argues, is science intensive and less capital intensive, so it may allow Kenya and other African countries to move forward under the guidance of national and regional policies on scientific and technological development and genetic conservation (Juma, 1989, p. 208).

The difference in Juma's views from Vandana Shiva's points comes in assessing modern science and the role of technology. While Shiva complains about the decontextualisation of modern science, Juma suggests that technological advances engineered by biotechnology should be placed in 'the broader context of scientific and technological development' which could enhance the capacity for African development (Juma, 1989, p. 208). Interestingly, he also rejects the idea of a simple transplantation of Western knowledge to Africa. He argues that neither neo-liberal economics, which focuses on the efficiency of resource allocation and does not take into account 'the main sources of economic change: genetic resources, technological innovation and institutional reform', nor Weberian rationality for institutional development could work in Africa (Juma, 1989, p. 210). He also refers to so-called village mentality, which is criticised for inhibiting the search for new knowledge. In his vision for African development, Juma argues for diversity, autonomy and the capacity to undertake experiments and relates this to both the scientific and social sphere (Juma, 1989, p. 210).

Juma advises a re-examination of the African 'culture in view of the imperatives of scientific and technological advances'. Some features of the African lifestyle, such as decentralisation of the community, can be useful, as it 'may lead to alternative technologies that are amenable to popular control'. So, by no means is he rejecting indigenous knowledge; instead, he calls for 'a viable mode of integration that leads neither to the erosion of the useful features of social organization nor inhibits the application of science and technology to social development'. He points out that replacing traditional practices with modern science is very costly and that Africa needs 'an alternative view of science', which would be based on neither 'the expansionist view of Bacon, nor the reductionist approaches of Descartes and the static notions of Newton'. Thus, African NGOs promoting science as the American Association for the Advancement of Science (AAAS) should serve as a laboratory for elaborating upon alternative science (Juma, 1989, p. 217). It should embrace the understanding that African agriculture relies on women's labour: another point in common with Vandana Shiva (Juma, 1989, p. 217). Such an integrated version of science should build links with local communities which are minimal at the moment and result in the public's distrust of science. A closer link between scientists and farmers should enable science to understand better the needs of the community and educate community groups (Juma, 1989, p. 218). His ideal of science sounds democratic in that the public can influence the direction of research and control decentralised production systems (Juma, 1989, p. 218).

In his model of development, Africa should create and multiply innovation clusters around key institutions such as universities and venture capital firms, particularly in the area of the life sciences and agriculture. This would bring knowledge spillovers, connect farmers with other institutions within the agricultural industry, increase agricultural productivity and improve African competitiveness in global markets. Transgenic agriculture is seen as an innovative one which could benefit African farmers, including smallholders (Juma, 2011). On the contrary, Vandana Shiva sees the agricultural industrial complex leading to hunger and malnourishment as a result of the promotion of monocultures and biofuels that are 'a greenhouse threat', the dependency on oil and GMOs as a solution to food security (Shiva, 2011).

Another area where Dr Shiva and Prof. Juma have different opinions is the interaction of developing countries with their former colonisers. Both were involved in the creation of international agreements on biodiversity. Juma was appointed Executive Secretary of the UN Convention on Biological Diversity while Shiva was 'appointed to an expert group to evolve the framework by the United Nations environment programme to implement Article 19.3 of the UN Convention on Biological Diversity' (Shiva, 2014). They each pointed out the dominant paternalistic role of institutional actors from Europe and the US within bio-diplomacy (Sanchez and Juma, 1994; Shiva, 2014). Shiva accuses the international corporations of the West, such as Monsanto, of establishing a monopoly empire over seeds and food (Shiva, 2014). For Juma, the responsibility of delaying the African biotech revolution is placed on environmental advocacy groups supported by the West and the European diplomatic corpus for imposing a restrictive approach on Africa in terms of the acceptance of GM crops (Juma, 2011). Biotechnology and transgenic agriculture are central in his strategy to overcome the dependency of the colonial age; he hopes that a '[N]ew age of biology – both as a field of scientific endeavour and as a metaphor on how we view the world' could bring changes to African development and can give Africa the chance to become 'an important player in the global knowledge ecology' (Juma, 2013, p. 14).

They differ in their perceptions of the roles of international research centres. Prof. Juma complains of the unfair transfer of technologies: '[T]here are numerous cases where machinery is acquired from other Third World countries and sold to Africa through European firms, often at exorbitant prices' (1989, p. 212). In certain cases, technological knowledge remains undisclosed. With fragmented and small national markets, African countries have little bargaining power. While he sees international research centres as an opportunity for developing countries to upgrade their innovative base, receive much needed capital from the West and improve bargaining power in international diplomacy, Shiva views them as agents of dominance from the West (Shiva, 1991).

The same event, the Zambian famine of 2002 and its treatment by the international community, is perceived by both authors in a totally different light. In 2001–2002, a severe drought took place in several African countries including Zambia. Its authorities rejected GM maize because they were worried that GM food would be used as part of the sowing of crops and that this would jeopardise European organic market possibilities (Paarlberg, 2008).

Juma disagreed with the position of the Zambian government on the basis of his conviction that GM crops produce a safe and nutritious food that people could eat, thus his belief that transgenic crops could help to prevent hunger in the future (Juma, 2011). Vandana Shiva, on the contrary, called sending GM crops to Africa 'inhuman aid', approved of the actions taken by the Zambian leadership and used this case to argue against Bt cotton, the only GM crop approved in India for commercial use (Shiva, 2003).

The GM banana campaigns

Banana is a staple crop for Africa. As with other crops, banana trees suffer from pests and diseases. The damage from the black Sigatoka fungus has been a problem since the 1970s and has reduced banana tree productivity by 40%. Technology has been employed to solve the issue; for example, tissue culture technology has been in use in East Africa since the mid-1990s (Juma, 2011, pp. 35–36).

Uganda has numerous initiatives to develop a transgenic banana that could be resistant to pests and diseases. Its government funded research to develop a GM banana resistant to black Sigatoka, nematodes and weevils (De Vries and Toennissen, 2001). In parallel, international Western-based charities also funded research on the GM banana, which resulted in protest activities from NGOs based in Uganda and outside. The rhetoric used in the anti- and pro-GM agenda has numerous references to the post-colonial discourse of former colonies in the developing world trying to cut off dependency from developed countries.

The Bill and Melinda Gates Foundation since 2005 has considered banana to be a staple crop along with cassava, rice and sorghum which could be enhanced for better micronutrient supply in the diets of the world's poorest populations. Making bananas less susceptible to disease was another goal of the project. Funding came under the Grand Challenges in Global Health Initiative to support genetic research by James Dale based in Queensland University of Technology (QUT), Australia and Ugandan scientists from the National Agricultural Research Organisation (Gates, 2012). The project has aimed to introduce Vitamin A and iron into the banana and is sometimes referred to as Golden banana due to the colour of the beta-carotene. Uganda has been chosen for a reason: banana is a staple food and the country has a wide range of local varieties, and several international research facilities are stationed there.

From 2001, Ugandan banana plantations have been suffering from banana Xanthomonas wilt (BXW) which was originally found in Ethiopia and cost East and Central African farmers half a billion American dollars per year (Nordling, 2010). It is no surprise that a BXW-resistant banana became another target for research in Uganda. The funding was provided by USAID and the Gatsby Foundation.

The BXW became a joint initiative by the African Agricultural Technology Foundation (AATF), the International Institute of Tropical Agriculture (IITA) and the National Agricultural Research Organisation (NARO). AATF negotiated for the royalty-free use of two genes from Academia Sinica in Taiwan, and by 2010 the project had reached the confined field trial phase (Ndwiga, 2012). Scientists from NARO and IITA based in Uganda developed 65 BXW-resistant lines which were planted in Kawanda (central Uganda) and approved by the National Biosafety Committee (Tripathi et al., 2014). The stages beyond the trial period are difficult to manage since the country lacks a regulatory framework and requires an amendment to the current Ugandan law which allows research only on GM crops and not commercialisation. The Biotechnology and Biosafety Bill was proposed and became a point of anti-GM activists blocking the initiative and entering the debate on the use of GM banana in Africa. Human trials on GM crops are not legally possible in the country (Ongu, 2015).

Surveys among local consumers indicated that the majority of Ugandans (58%) were willing to consume a GM banana (Kikulwe et al., 2011; Ongu, 2015). The authorities admitted that Uganda follows Europe in its cautious approach to GMOs. The country is a part of the Cartagena Protocol (2001). The reasons for such caution relate to the limitations of institutional capacity, such as the small number of national scientists adequately trained, and the cost of equipment, so decisions about GM crops are difficult to implement (Mugwanya Zawedde, 2014).

The debate over the GM banana in Uganda quickly moved beyond national borders, receiving attention from American, Canadian and Indian activists. The Ugandan anti-GM organisations have included the Ugandan branch of the UK-based NGO ActionAid and Alliance for Food Sovereignty in Africa (AFSA) which is a 'Pan African platform comprising networks and farmer organizations working in Africa' (AFSA, 2016). While most NGOs, members of the network, are African, the network also includes Global Justice Now. This NGO, formerly known as Third World Movement, based in the UK is dedicated to 'fighting injustice, particularly in the global south' (GJN, 2016a). In 2016, it produced a report entitled 'Gated Development' in which it criticised Bill Gates and his foundation for supporting corporate business and funding research into GM crops and claims that it is funding organisations to push GM crops across Africa. It calls for a change in national legislation on this issue (GJN, 2016b).

Two Ugandan activists supporting the GM banana are Isaac Ongu and the Science Foundation for Livelihoods and Development. Ongu has published a number of media articles in the online journal of the Genetic Literacy Project in favour of GM banana, arguing that this staple crop is essential for Africa and 'biotechnology becomes a necessity and not just a debate topic for the privileged class in the west' (Ongu, 2015).

In May 2013, Navdanya, Vandana Shiva's organisation, launched a campaign 'No to GMO bananas'. First, the Gates Foundation's bio-fortified banana was perceived as a threat to India because of its relevance to two new laws proposed by the Indian government to deregulate GMOs for food and agriculture which could be called India's Monsanto Protection Act and the Food Security Act which supports the bio-fortification of crops in India (Navdanya, 2016). The organisation wrote a 'GMO banana petition' to the Indian prime minister, asking him not to follow up on the GM banana in India. The 'No to GMO bananas' campaign was accompanied by another campaign called 'Seeds of Freedom, Gardens of Hope', which offered openly pollinated seeds from Navdanya as a GMO-free, iron-rich, nutrient-rich food alternative. While Dr Shiva was on a visit to East Java in 2014, she met with farmers from the Kediri region who served her yellow and red bananas, which inspired her to invite activists to research on Vitamin A-rich indigenous bananas and discover from where the developers of GMO bananas got their Vitamin A traits. Later that year, Navdanya launched the campaign internationally under the name 'No GMO Banana Republic – Stop Banana Biopiracy' at the Bhoomi Festival in New Delhi. The campaign included the issue of An Open Letter to QUT's Dr James Dale, the Bill and Melinda Gates Foundation and the Convention on Biological Diversity (Seed Freedom, 2015). The campaign song 'We don't want no Pirate Banana' was written and translated into English, Hindi, Spanish and Swahili.

Next, Dr Shiva connected with graduate students at Iowa State University who had also held a protest against human trials of GM bananas on student volunteers. Six students organise a silent protest, wore gas masks and banana costumes to spread the message: 'Support Food Democracy'. Shiva travelled to Iowa and delivered a lecture to a student audience in support of their protest (Seed Freedom, 2015).

In April 2016, Dr Shiva arrived in Brisbane, Australia, to deliver the AFSA's Open Letter, signed by 57,000 supporters, about the Gates Foundation-funded GMO bananas that had been developed at QUT and tested on female students at the Iowa State University.

A Seattle-based NGO, 'Community Alliance for Global Justice' (CAGJ), launched a campaign called AGRA Watch to challenge the 'Bill and Melinda Gates Foundation's questionable agricultural programmes in Africa, including its Alliance for a Green Revolution in Africa (AGRA)'. AGRA was founded in 2006 through a partnership between the Rockefeller Foundation and the Bill and Melinda Gates Foundation. It supports effective public-private partnerships for improving smallholder farming in Africa and has a 'Strategy for an African Green Revolution' which is based on knowledge and technology spread in African countries. In the case of the GM banana, it supports the initiative under which Uganda would allow the commercialisation of GM crops (AGRA, 2016). AGRA Watch investigates AGRA, its projects and funding and has accused AGRA of promoting 'a form of philanthrocapitalism based on biopiracy', where bio-piracy is understood as the legal mechanisms which allow the global North to access southern genetic resources (CAGJ, 2016). Its website refers to the same 57,000 signatures collected in protest against GM banana trials, the Iowa students' network and an open letter submitted to the Gates Foundation. In February 2016, its members rallied outside the Gates Foundation office in Seattle for an hour, with three members wearing banana costumes and carrying signs demanding an end to the funding of research on GMOs (AGRA Watch, 2016).

Literature

AFSA (2017) What is AFSA. http://afsafrica.org/what-is-afsa/ as viewed 02.05.2017.

AGRA (2016) Who are we? http://agra.org/who-we-are/ as viewed 05.05.2016.

AGRA Watch (2016) AGRA watch action report. https://agrawatch.wordpress.com/ as viewed 06.05.2016.

Allen, K. (2013) Meet Calestous Juma, Africa's genetically modified crop 'optimist'. *The Star*, 17 June 2013. www.thestar.com/news/world/2013/06/17/meet_calestous_juma_africas_genetically_modified_crop_optimist.html as viewed 06.05.2016.

Bhatt, V. (2015) Interview with Dr Vinod Bhatt. Dehradun area, 28 February 2015.

Boesen, J., Storgaard Madsen, B., Moody, T. (1977) *Ujamaa: Socialism from above*. Uppsala: Scandinavian Institute of African Studies.

CAGJ (2016) AGRA watch. https://cagj.org/agra-watch/ as viewed 06.05.2016.

The Chopra Foundation (2016) Speakers: Vandana Shiva. www.choprafoundation.org/speakers/vandana-shiva/ as viewed 02.05.2017.

De Vries, J., Toennissen, G. (2001) *Securing the harvest: Biotechnology, breeding and seed system for African crops*. Wallingford: CABI.

Gandhi, M.K. (1921) "Hind swaraj" or the "Indian home Rule", *Young India*, 22 January 1921.

Gandhi, M.K. (1995) *A golden treasury of wisdom: Thoughts and glimpses of life*. Mumbai: India Printing Works.

Gandhi, M.K. (1997) *Hind Swaraj and other writings Mohandas Gandhi*. Ed. A.J. Parel. Cambridge: Cambridge University Press.

Gandhi, M.K. (2016) Gandhi's views on environment nature cure & holistic treatment. Mani Bhavan Gandhi Sangrahalaya: Mahatma Gandhi Information Website. www.gandhi-manibhavan.org/gandhiphilosophy/philosophy_environment_naturecure.htm as viewed 06.05.2015.

Gates, W. (2012) A bunch of reasons: Building better bananas. htpps://gatesnotes.com/Development/Building-Better-Bananas as viewed 08.06.2016.

Howard, L.E. (1953) *Sir Albert Howard in India*. London: Faber and Faber.

GJN (2016a) About us. http://www.globaljustice.org.uk/about-us as viewed 02.05.2016,

GJN (2016b) Gated development: Is the Gates Foundation always a force for good? www.globaljustice.org.uk/sites/default/files/files/resources/gjn_gates_report_june_2016_web_final_version_2.pdf as viewed 02.05.2017

Jalees, K. (2014) Reply to the article by Prof. Calestous Juma 'Feeding Africa': Why biotechnology sceptics are wrong to dismiss GM. Navdanya blog entry. www.navdanya.org/blog/?p=1427 as viewed 05.06.2014.

Juma, C. (1989) *The gene hunters: Biotechnology and the scramble for seeds*. Princeton: Princeton University Press.

Juma, C. (2011) *The new harvest: Agricultural innovation in Africa*. Oxford: Oxford University Press.

Juma, C. (2012) Curriculum Vitae. November 2012. https://apps.hks.harvard.edu/faculty/cv/calestousjuma.pdf

Juma, C. (2013) Biosciences in Africa's economic transformation, in B. Heap, D. Bennett (eds.) *Insights: Africa's future . . . Can biosciences contribute?* Cambridge: Lavenham Press, pp. 11–15.

Kikulwe, E.M., Wesseler, J., Falck-Zepeda, J. (2011) Attitudes, perceptions, and trust: Insights from a consumer survey regarding genetically modified banana in Uganda. *Appetite*, 57(2): 401–413.

Kranz, L. (2015) Harvard professor failed to disclose connection. *Boston Globe*, 1 October 2015. www.bostonglobe.com/metro/2015/10/01/harvard-professor-failed-disclose-monsanto-connection-paper-touting-gmos/lLJipJQmI5WKS6RAgQbnrN/story.html as viewed 06.05.2015.

London, S. (2016) In the footsteps of Gandhi: An interview with Vandana Shiva. www.scottlondon.com/interviews/shiva.html as viewed 06.05.2016.

Masefield, G.B. (1950) *A short history of agriculture in the British colonies*. Oxford: Clarendon Press.

Mbembe, A. (2001) *On the postcolony*. Berkley: University of California Press.

Mugwanya Zawedde, B. (2014) Challenges and solutions for implementing risk assessment and risk management at the research stage in Africa: The case of Uganda. 13th ISB-GMO Symposium, Cape Town. 11 November 2014.

Navdanya (2016) Navdanya launches No to GMO bananas campaign. www.navdanya.org/news/338-navdanya-launches-no-to-gmo-bananas-campaign as viewed 05.05.2016.

Nayar, P.K. (2012) *Colonial voices the discourses of empire*. Chichester: Wiley-Blackwell.

Ndwiga, J. (2012) *Developing a bacterial wilt resistant banana for smallholder farmers in Africa*. Nairobi: AATF. http://aatf-africa.org/Developing-bacterial-wilt-resistant-banana as viewed 06.05.2016.

Norlding, L. (2010) Uganda prepares to plant transgenic bananas. *Nature News*. 1 October 2010. www.nature.com/news/2010/101001/full/news.2010.509.html

Nyerere, J.K. (1968) *Freedom a socialism: Uhuru na Ujamaa*. Nairobi: Oxford University Press.

Ongu, I. (2015) As disease threatens Uganda's banana crop, ActionAid and anti-GMO groups fan fears. Genetic Literacy Project. 27 July 2015. www.geneticliteracyproject.org/2015/07/27/as-disease-threatens-ugandas-banana-crop-actionaid-and-anti-gmo-groups-fan-fears/ as viewed 06.05.2016.

Paarlberg, R. (2008) *Starved for science: How biotechnology is being kept out of Africa*. Cambridge: Harvard University Press.

Pearce, F. (1991) *Green warriors: The people and the politics behind the environmental revolution*. London: The Bodley Head.

Peritore, P. (1999) *Third world environmentalism case studies from the Global South*. Gainesville: University Press of Florida.

Pinto, V. (1998) *Gandhi's vision and values: The moral quest for change in Indian agriculture*. New Delhi: Sage.

Popham, P. (2014) Meet Vandana Shiva: The deserving heir to Mahatma Gandhi's legacy: What Gandhi was to the British empire, Dr Shiva is to Monsanto. *The Independent*, 20 August 2014. www.independent.co.uk/voices/comment/meet-vandana-shiva-the-deserving-heir-to-mahatma-gandhis-legacy-9681770.html as viewed 06.05.2016.

Saigal, N. (2013) India after Gandhi: From the British Raj to Monsanto Raj. *The Alternatives*, 30 August 2013. www.thealternative.in/society/india-after-gandhi-monsanto-raj-bt-cotton/ as viewed 07.05.2016.

Sanchez, V., Juma, C. (1994) *Biodiplomacy: Genetic resources and international relations*. Nairobi: African Centre for Technology Studies.

Seed Freedom (2015) GMO piracy – of banana – connecting dots. http://seedfreedom.info/gmo-biopiracy-of-banana-connecting-the-dots/ as viewed 06.05.2016.

Shiva, V. (1991) *The violence of the green revolution: Third world agriculture, ecology and politics*. London: Zed Books.

Shiva, V. (1993) *Monocultures of the mind: Perspectives on biodiversity and biotechnology*. London: Zed Books.

Shiva, V. (2000) *Stolen harvest: The hijacking of the global food supply*. Cambridge: South End Press.

Shiva, V. (2001) *Protect or plunder?: Understanding intellectual property rights*. London: Zed Books.

Shiva, V. (2003) Why I believe that sending GMOs to hungry people is 'inhuman aid', HAR, Summer 2003. http://stopogm.net/sites/stopogm.net/files/InhumanAid.pdf as viewed 06.05.2016.

Shiva, V. (2005) *Earth democracy: Justice, sustainability, and peace*. Cambridge: South End Press.

Shiva, V. (2011) The agricultural industrial complex, in S. Best, R. Kahn, A. Nocella II, P. MacLaren (eds.) *The global industrial complex: Systems of domination*. Lanham: Lexington Books, pp. 169–195.

Shiva, V. (2012) *Making peace with the Earth: Beyond resources, land and food wars*. New Delhi: Women Unlimited.

Shiva, V. (2014) Fine prints of the food wars. *Countercurrents*. www.countercurrents.org/shiva030814.htm as viewed 06.05.2015.

Shiva, V. (2016) Dr Vandana Shiva about. http://vandanashiva.com/?page_id=2 as viewed 06.05.2016.

Specter, M. (2014) Seeds of doubt: An activist's controversial crusade against genetically modified crops. *New Yorker*, 25 August 2014. www.newyorker.com/magazine/2014/08/25/seeds-of-doubt as viewed 06.05.2016.

Thirtle, C., Piesse, J., Gouse, M. (2005) Agricultural technology, productivity and employment: Policies for poverty reduction. *Agrekon*, 44(1): 37–59. https://core.ac.uk/download/files/153/6557388.pdf as viewed 07.05.2016.

Tripathi, L., Nath Tripathi, J., Kiggundu, A., Korie, S., Snotkoski, F., Tushemereiwe, W.K. (2014) Field trial of Xanthomonas wilt disease-resistant bananas in East Africa, 32(9): 868–870.

von Freyhold, M. (1979) *Ujamaa villages in Tanzania: Analysis of a social experiment.* London: Heinemann.

Wallace, T., Martin, J.T. (1954) *Insecticides and colonial agricultural development.* London: Butterworths.

4   Regionalism, food sovereignty and GM crops

Introduction

Regionalism is a traditional subject of international relations studies. The concept of 'region' is 'an ambiguous term'; it has more than a geographical component and definitely includes discussions of security issues (Nye, 1968, p. VI). Region can have a geographical meaning with proximity and specificity being key criteria and strategic and functional meaning which includes economic, environmental and cultural commonality (Vayrynen, 2003, p. 26). Physical definitions of regions are attempts from states to reaffirm their boundaries and organise their territories, while 'functional conceptualizations of regions emanate from the interplay of subnational and transnational' processes which are only partially controllable (Vayrynen, 2003, p. 27). The politics of regionalism is a response to the larger processes of globalisation and particularly its aspect of international trade, economic and political integration. From this perspective, states may perceive regionalism as a defence mechanism against rising international competition (Pelagidis and Papasotiriou, 2002). Thus, regionalism is an accompanying process of globalisation. It can be a strategy of a state or region to adapt to new challenges in the international environment and provide some degree of autonomy. Castells suggested the notion of a regionalised global economy: 'That is, a global system of trade between trading areas, with increasing homogenization of customs within areas, with increasing homogenization of customs within the area, while maintaining trade barriers vis-à-vis the rest of the world' (Castells, 2000, p. 111).

Castells's concept of networks is also relevant to discussions on regionalism. In the globalised trade system, actual trading units include firms, multinational companies (MNCs) and their networks (Castells, 2000, p. 115). At the same time regions 'under the impulse of their governments and business elites, have restructured themselves to compete in the global economy, and . . . have established networks of cooperation between regional institutions and between region-based companies' (Castells, 2000, p. 412). Regional authorities are left by their states to connect with the local community and can also develop strategies to compete in the global system under the processes of larger integration and regional decentralisation. To link up, they need to (re)construct a particular identity (ethnic, territorial, religious) which may take over that of the state (Castells, 2004, p. 339). This point is in line with Kollmorgen's definition of regionalism as a

> clustering of environmental, economic, social, and governmental factors to such an extent that a distinct consciousness of separate identity with the whole, a need for autonomous planning, a manifestation of cultural peculiarities, and a desire for administrative freedom, are theoretically recognized and put in practice.
>
> (Kollmorgen, 1945, p. 385)

Agriculture has always been a sensitive area for national governments. Since the establishment of the Bretton Woods regime in international trade, international organisations have pushed for more liberalisation in trade, while national governments have resisted. Trade in agricultural products was discussed at different rounds of GATT (the General Agreement on Tariffs and Trade), such as the Dillon Round (1960–1961) and the Kennedy Round (1963–1967), without much success in reducing trade barriers and tariffs. Only after the launch of the Uruguay Round (1986–1994) did trade liberalisation in agriculture begin to move forward (Lee, 2007). Thus, since the mid-1990s, national governments, farmers and their organisations have been concerned with new changes and have had to consider possible responses to protect their interests.

Another important concept to be discussed in relation to the discourse of regionalism is food sovereignty. This is a derivative of the notion of food security which places the focus on self-sufficiency in food production and at household, community, regional and state levels. The first conceptualisation of food sovereignty dates back to 1996 when La Via Campesina and the Food and Agriculture Organization of the United Nations (FAO) introduced food sovereignty in their position papers. The FAO held the World Food Summit, and in its action plan specified that 'each State must adopt a strategy consistent with its resources and capacities to achieve its individual goals . . . and, at the same time, cooperate regionally internationally in order to organize collective solutions to global issues of food security' (FAO, 1996).

La Via Campesina, which is now an international movement – a network of NGOs working and representing farmers – was set up in 1993 in Mons, Belgium. It has since grown to represent about 200 million farmers through 164 local and national organisations in 73 countries from Africa, Asia, Europe and the Americas (La Via Campesina, 2011). It defines food sovereignty as 'the right of each nation to maintain and develop its own capacity to produce its basic foods respecting cultural and productive diversity' (La Via Campesina, 1996). Priority is given to 'local food production and consumption'. Such an interpretation of the concept 'gives a country the right to protect its local producers from cheap imports and to control production' (ARC, 2014).

Both food security and food sovereignty refer to food production and consumption, yet these are different. Some have argued that 'food security' is a technical term while 'food sovereignty' is a political term. For example, FAO experts argued:

> The concept of food security – adopted by FAO member states – is somehow a neutral concept in terms of power relations. It does not prejudge the concentration of economic power in the different links of the food chain and in the international food trade, or the ownership of key means of production such as land, or more contemporarily, access to information.
>
> (Gordillo and Mendez Jeronimo, 2013, p. vi)

Another means of production in agriculture is seeds. Under the global trend towards the liberalisation of international trade, states have to dismantle protective mechanisms at national level. National genetic resources and biotech innovations have been seen as trade-related aspects of intellectual property rights. This has effectively led national governments to be obliged to uphold patent protection over the microorganisms and microbiological processes used in agricultural production, as presented in the International Treaty on Plant Genetic Resources for Food and Agriculture (Lee, 2007). This has become a matter of concern in both developing and developed countries, particularly those in which the agricultural sector is less competitive in global terms. So 'the concept of food sovereignty begins precisely with noting the asymmetry of power in the various markets involved and the various spheres of power involved in food, as well as in the areas of multilateral trade negotiations' (Gordillo and Mendez Jeronimo, 2013, p. vi).

On this ground, it is possible to argue that while some NGOs push for food sovereignty from the perspective of farmers, national governments interpret the concept in their agenda at regional level in political terms. This argument can be illustrated with two case studies of NGOs opposing GM crops – Danube Soya and National Association for Genetic Safety (NAGS).

GENET, GMO-free Europe and Danube Soya

For a quick pre-Christmas get-together dinner, I invited my Bulgarian and German friends. The German is a devoted vegetarian and volunteered to bring a dessert. It is a mix of frozen berries and soya yoghurt which tasted delicious. For my other friend and myself, soya yoghurt is a novel idea. My next soya experience occurred in Germany when teaching in Berlin. In the German capital's food shops, almost everything is organic and bio-labelled, and I could not find a small tin of cow's milk in any city centre grocery shop, so I ended up adding soya milk to my tea for breakfast. It tastes fine, but it does require the palate to get used to it.

My first encounter with organic soya was during the visit to the GMO-free Europe Conference in 2015. But, first, a little bit of background. The GMO-free Europe, a network of NGOs lobbying against GM crops, is a spin-off organisation by GENET, which is a 'European network of non-governmental non-profit organizations engaged in the critical debate of genetic engineering, founded in 1995 in Switzerland'. Its mission is 'to provide information on genetic engineering to its member organizations and the interested public and to support their activities and campaigns'. Currently it has 51 organisations listed as members in 27 European countries (GENET, 2016, para. 1). It can be argued that both networks represent a regional discourse in the GM crop debate and serve as examples of how civil society groups cooperate with regional and municipal authorities.

GMO-free Europe has been run as a series of conferences. The first listed on its website is one held in Vienna in 2003 'to support the strategic and practical work of civil society groups to foster those agricultural and food processing practices which do not need GE and to strengthen the groups to resist the continuing pressure to adopt GE in agriculture' (GENET, 2003).

In 2005, the conference was held in Berlin and attended by 250 delegates. Among them a group of representatives of European national and municipal authorities, including Andrä Rupprechter from the Ministry of Agriculture, Austria (which then held the EU presidency); Josef Stockinger, Agricultural Minister of Upper-Austria; Fabio Boscaleri, Agricultural Ministry of Tuscany, Italy; Pawel Polanecki, Deputy Chairman of the Sejim of Mazowie, Poland; Gabriele Friderich, City Council of Munich, Germany; Janusz Wojciechowski, Vice-President of the European Parliament's Committee for Agriculture, Poland; and Hansjörg Walter, Member of the Swiss Parliament and President of the Farmers' Union of Switzerland (GMO-free Europe, 2005b). The conference issued the Berlin Manifesto for GMO-free Regions and Biodiversity in Europe. The document proclaimed the right and duty of regions to decide 'about the use of genetically modified organisms (GMOs) and the shape of our landscape' and stressed that such decisions were not to be imposed by individual farmers, bureaucrats or companies (GMO-free Europe, 2005a). Another right and duty given to the regions was to protect seeds from genetic modification in order to remain what can effectively be called regional food sovereignty:

> Protecting and encouraging the conservation and breeding of native and adapted local varieties and the integrity of farm saved seeds is an important duty and right of regional agricultural policy. As seeds reproduce there can be no thresholds for any unlabelled GM contamination of non-GM conventional, organic and traditional varieties.
>
> (GMO-free Europe, 2005a)

After two years of silence (the previous conference was held in 2012), in 2015 the Group produced three (!) parallel conferences with the following agendas: the European GMO-free Regions Network, Danube Soya Association and NGOs and Scientists network of GMO-free regions. In the tradition of the 2005 Berlin conference of GMO-free Europe, the Danube Soya Association had prepared the culmination of the joint events – the signature of the Berlin Declaration on 8 May 2015 at the North Rhine-Westphalia Representation. The Declaration contains six articles and aims at complete withdrawal of GMO crops across Europe (GMO-free Europe, 2015a).

Perhaps a little explanation on the European regional agricultural context might be helpful in understanding the rationale for such an organisation as Danube Soya. Agriculture has always been a sensitive topic for the EU. On the one hand, member states must ensure there is enough food for their citizens, and in the immediate post-war years, such supplies were low. On the other hand, Europe increased its self-sufficiency in food production and soon other problems appeared – agricultural surpluses which demanded market interventions into supply and demand. To manage agricultural policies in common space, the European Community (EC) has created the Common Agricultural Policy (CAP). This is an example of a regional European agricultural policy which reflects the duality of EC member states' position: states had to compete among each other within the integrated European market and coordinate joint policy to sustain external competition. The weight of farming lobbies has been traditionally solid, which had to be acknowledged by politicians, and in addition, farmers' interests should be represented by numerous European NGOs and their networks. For example, the Agricultural and Rural Convention (ARC2020) was set up in 2010 as a multi-NGO platform and since 2013 it has evolved into a separate NGO whose main activities are aimed at 'offering a space to regenerate public debate in Europe, while also drawing the public's attention to both controversies and narratives on future farming, food and rural policies' (ARC, 2016). So it appears to be a normal European practice to set up an NGO to lobby on behalf of farmers' interests.

Agricultural production in developed countries has contributed to the increase in policies, but world economic activity has caused a reduction in demand and, in turn, the prices for many crops such as cereals have stagnated. Technological progress has allowed farmers to produce more at less cost, promising to produce even more agricultural supply. In light of further technological advances, particularly in biotechnology, Europe and America have experienced enormous pressure to find uses for thousands of tonnes of excess produce. For example, this explains why high fructose corn syrup has been developed and added to numerous food products. It is clearly related to the existence of an unsaleable maize surplus in the US (Raikes, 1988).

It is a similar story with soya. On the one hand, soya beans have resulted in a decrease in the production of animal feed and partially substitute cereals, while on the other, they have also provided new opportunities for market development across the whole production chain. This has been reflected at international trade negotiations.

In the Dillon Round of GATT, the EC entered into oilseed binding. By that time, European animal feed compounders found that a mixture of carbohydrates and vegetable protein were an acceptable and much cheaper substitute for the cereals used in animal feed. As a result of this demand, the supply of soya beans has grown, and it has affected the whole value chain of soya meal production, from growers to bean crushers. Yet, in the 1970s, the EC introduced a support system for oilseeds under its CAP, which paid direct subsidies to European soya crushers. When soya bean prices fell, the subsidies were increased and vice versa in order that European soya growers could maintain the same prices (Swinbank and Tanner, 1996, p. 103).

The German Farmers' Union, the Deutscher Bauernverband, has been an important political player in German politics due to its links to the ruling political alliance of the Christian Democratic Union of Germany and Christian Social Union in Bavaria. For decades, Deutscher Bauernverband promoted the ideal of a family farm as the main goal of agricultural policy and supported producers' prices so that each farming family could afford an average national standard of living (von Cramon-Taubadel, 2000, p. 412). Since joining the EC, German agriculture was the least competitive. With the country having lost some of its agricultural land due to its partition, crop yields were lower than in other member states and its protectionist measures were the most expensive, since German farmers were paid the highest cereal prices in Europe (Meunier, 2005; Ackrill, 2000).

In the 1990s, Germany faced several challenges. With reunification, the country had to restructure its farming sector. As an EC member, it had to participate in the Uruguay Round of GATT and accept the CAP reform, which was not welcomed by the German farming lobby (Wilson and Wilson, 2001; Von Cramon-Taubadel, 2000). The European expansion eastward also meant additional burdens to the CAP budget with the potential to pressurise agricultural prices in the European market and add to the threat of further agricultural surpluses (Wilson and Wilson, 2001, p. 262).

Danube Soya represents this trend to provide better opportunities to European farmers and develop a regional market as a countermeasure to external dependency in food production. While it was initially developed in Austria, the initiative has full support from the German authorities because it matches the German strategy to support farming families and address the current challenges caused by the European extension eastwards and the possible dismantling of CAP. The initiative is supported by funds from Austrian and German development agencies – Deutsche Gesellschaft für Internationale Zusammenarbeit and the Austrian Development Cooperation (Danube Soya, 2016). The meetings of the organisation's conference in 2015 were held in the Austrian Embassy and the Representation Offices of Thuringia, Hesse and the North Rhine-Westphalia in Berlin. The Austrian Embassy hosted a generous evening reception for all participants of the GMO-Free Conference. Dr Nikolaus Marschik, the Austrian Ambassador to Germany, delivered a welcome speech, which was followed by the symbolic act of planting soya seeds by Max Hiegelsberger (Regional Ministry for Agriculture Upper Austria), Horst Becker (Parliamentary State Secretary Ministry of Climate Protection, Environment, Agriculture, Conservation, and Consumer Protection of the State of North Rhine-Westphalia) and Matthias Krön, President of Danube Soya (GMO-free Europe, 2015b).

After the first meeting in 2010 and the initial 'dream', the project was developed into an association in 2012 (Gaugitsch, 2015). It was registered in Vienna, Austria, as an international non-profit association, or NGO. Its chairman, Matthias Krön, said in 2015 that it is 'a public organization, not a lobby'. The aim of the association is 'to bring farmers, seed and feed companies, traders, processors and animal owners on board with the same concept – promotion of the European protein supply', but it must be GMO-free. As seen from the name, the association promotes the cultivation of GM-free soya in Europe. Indeed, in Krön's own words, soya 'should receive a European passport' (2015b).

The name 'Danube Soya' for the association is a smart one. The Danube (Donau) is one of the largest rivers in Europe and flows through Germany, Austria, Slovakia, Hungary, Croatia, Serbia, Romania, Moldova, Ukraine, and Bulgaria. Its tributaries, however, also include Bosnia and Herzegovina, the Czech Republic, Slovenia, Montenegro, Switzerland, Italy, Poland, the Republic of Macedonia and Albania. And these countries, the main targets of the association, make up a good chunk of EU members and associated countries. In addition to being a common trade region with shared identities, the Danube region also composes half of Europe's arable land. Some of these countries are more developed than the others, and the leadership of Germany is obvious. This is confirmed in the fact that German companies, members of the association, are acquiring or interested in gaining access to agricultural lands in the Danube countries to grow GM-free soya. This is, of course, not a coincidence. It makes Danube Soya a good name and a smart historic allusion.

In the project description, Danube Soya leaders begin with the presentation of soya as 'a challenge', specifically the North American export of soya to Europe: 'Our dependence on soya import thus becomes a challenge for all of Europe' (Danube Soya, 2016). Thus, the initiative seeks to address this challenge as 'it provides a foundation for the production of high quality, origin-controlled GMO-free food and feed for the Danube region and the Western European market'. Further, it 'will constitute a significant contribution to the independence of the European protein supply' (Danube Soya, 2016).

So why does the Association focus on soya? Because soya represents protein supply and creates new market opportunities. On average, the European population consumes 105g of protein per capita per day, 59% of which is from animal-derived products. The European Union imports soya beans and meal in large quantities. For example, 28 million tonnes were imported in 2013 (De Visser et al., 2014). This low self-sufficiency in protein-rich feed is disturbing and has been discussed in the European parliament (Hausling, 2011). It is Europe's 'protein dependency' on countries such as Brazil, Argentina and the United States which poses an economical, ecological and political challenge, given the fact that around 95% of imported soya is produced with genetically modified seeds. Moreover, the demand for soya is constantly growing. Danube Soya (2015) suggested that European soya bean cultivation could increase by 300% within the next five years. In the session and B2B meeting of the Association in April 2016, soya producers were introduced to buyers and the audience was informed about the potential for the soya market to be developed beyond the feed market. This includes lecithin, tofu and soya milk production as well as products developed for the vegetarian diet, such as soya substitute products for meat (Danube Soya, 2016).

Partnering through the Association allows Western European companies to relocate their supply chain to Europe and obtain control over soya producers. The major rule is that soya crops are not GM. At the conference in 2015, the audience was reminded of the importance of keeping links between Western and Eastern Europe. In his speech, the chairman stressed that Ukraine, as a new addition to the association, has been very significant. In European terms, Ukraine has large, arable lands which are highly fertile due to its famous chernozem. According to Krön, soya import to Europe is a more difficult dependency than 'the dependency on oil and gas'. This makes Danube Soya and Ukraine an interesting case to be placed in the current context of tensions between Europe and Russia.

There is also a geographic pattern relating to the soya bean buyers and sellers invited by the Association to its meeting in 2016: the buyers came from Germany and Austria, while sellers came from Central and Eastern Europe – Czech Republic, Croatia, Moldova, Serbia and Ukraine (Danube Soya, 2016). An Austrian commercial laboratory which was jointly represented with an Austrian governmental institution, Environmental Agency Austria, offers consultancy services to the Danube Association's members. The move eastwards is also reflected in the fact that the Danube Soya Congress 2016 took place in Budapest, Hungary.

It is possible that both GM and non-GM soya can satisfy the growing European demand. On the basis of the quality of products and food safety, there is not much difference for consumers, unless they are taught to make preferences on other grounds; this is undertaken by explaining to European consumers the possible risks for human health and the environment from GM crops. But there is a difference in price. On the one hand, a European farmer will have to add €5 per pig fed with non-GM soya (Krön, 2015a). On the other, the cultivation of GM plants in Europe is subject to strict control and a regulatory system: only two crops are allowed (corn and potato), and in some European countries and regional communities, all GMO seeds are banned. In order to still enter the competition to grow the strategic crop and make the case for GM-free soya animal feed in the region despite its higher price, European countries need a strong political incentive. In this case, non-GM producers are supported by the Austrian and German state and regional authorities who have already promoted regulation on non-GM animal feed and sponsored voluntary GMO-free labelling for food companies. Both governments provide different support, including funding. So the association is more a political project than an act of civil society protest against GMO. This explains why European integration rhetoric is being used in the promotion of the Danube Soya Association.

The Danube Soya Association focuses a lot on business and political networking and labelling. Its activism includes a small number of public campaigns. These include an image campaign which 'creates[s] light illustrations that pick up the doubts and fears of soya production in Europe' (Behance, 2014). A series of watercolours has been produced by two Austrian artists, Brigitte Baldrian and Harald Hackel. The illustrations depict major issues of food security, such as CO_2 emissions, overpopulation, deforestation and the negative impact of agriculture on the environment. The word 'GMO' is placed in a yellow triangle and crossed out. These artworks do not require any textual explanation and have been used as visual material at 'some conferences in Europe' (Behance, 2014).

The style of this digital campaign has similarities with another series of illustrations developed for EuropaBio, the natural enemy of Danube Soya since it on the contrary promotes the use of transgenic agriculture in Europe. The video 'Bureaucratic Barriers to Biotech', made in 2012, tells the story of how biotech science has been blocked in Europe and uses the same EU symbols (EuropaBio, 2012).

Russia: WTO, food sanctions, GMOs and NAGS

Most Western readers are not familiar with Russian public debates on GM crops. Yet these debates reflect similar trends to those found in global debates on GM crops and are interesting examples of how GM crops have become a topic within the national agricultural policy.

At a glance, the official Russian position on GM crops appears to be negative. Yet there are many nuances, and the political context has been changing. In 2013, I acted as an expert, ranking start-up projects for Skolkovo, a state-funded innovation centre often called the 'Russian Silicon Valley'. At the start-up conference, I met several people including one businessman who was working on different projects, one of which was agricultural. In addition, he worked as a partner for a senior partner, a well-known international business figure. From brief conversations, it appeared that Russian businessmen were very willing to consider GM technology as a tool to increase crop yields and secure better profits. However, top international management was more cautious and perhaps even better informed.

The 'Food Security Doctrine of the Russian Federation' (hereafter, the Doctrine) was signed by then President Medvedev on 30th January 2010 and includes the formal position of the Russian authorities on food security. The structure of the Doctrine consists of the identification of risks towards food security and possible policy responses.

The duty of the state to guarantee food security for Russian citizens has been underlined in the document several times. Food security was named as one of the major elements of national security in the medium term. The policy goals listed in the Doctrine included 'reliable food supplies to the population, developing the national agro-industrial and fish-producing sectors, responding promptly to internal and external threats to the stability of the food market, and participating effectively in international cooperation in the field of food security' (Russian Federation, 2010, p. 1). Thus, food security has been understood as food independence, which includes a set of appropriate goals such as 'timely forecasting, detection and prevention of internal and external threats endangering food security, minimization of their negative consequences by ensuring constant operation of the system that supplies people with food' (Russian Federation, 2010, p. 3).

The challenges mentioned in the Doctrine include macroeconomic risks, resulting from the decreased attractiveness of Russian industry, the compatibility of national production and dependency on major factors of the economy on international economic institutions; technological risks, caused by the delay in the development of industrial innovation in Russia, differences in standards in food safety and the system of food safety control; agro-ecological risks associated with climate change and resulting from natural and man-made catastrophes; and international risks, caused by fluctuations in international markets and state protection measures in foreign countries (Russian Federation, 2010, pp. 7–8).

The Doctrine has captured the decline of the Russian agricultural sector, illustrating that major sectors, such as the dairy, fishery and meat industries could satisfy only 50–70% of domestic demand. This also resulted in food price doubling in the period 2005–2008 (Ushachev, 2008). Agriculture was in many ways, including in terms of innovation and institutional infrastructure, left behind after the collapse of the Soviet Union. The agriculturalist and politician Konstantin Babkin, in his opening speech at the Agriculture Section of the Moscow Economic Forum in November 2012, joked that Russians would die off just as the mammoths did unless the decay of rural development and degradation of agriculture are halted (Torgovo-Promyshlennaya Palata, 2012). Thus, one of the targets was to upgrade existing agricultural capacities to satisfy the national demand for food and provide a higher degree of independence from agricultural exports. This can partially explain why Skolkovo businessmen were considering GM technology in 2013.

Another important event was Russia's joining the World Trade Organization (WTO). The Working Party on the Accession of the Russian Federation was established in June 1993 and completed its mandate in November 2011. The Eighth Ministerial Conference approved the Accession Package of the Russian Federation on December 2011. On 22 August 2012, the Russian Federation became the WTO's 156th member (WTO, 2012). Of course, joining the WTO with a less competitive agricultural industry would have been a concern for the Russian government and it also seemed to naturally coincide with the food sanctions, which are a modern example of protectionist policy in agriculture, in 2014.

In 2014, the EU imposed political sanctions on Russia as a result of the conflict in Ukraine. In response, the Russian government also imposed sanctions on Europe, restricting agricultural imports from EU member states. On the one hand, sanctions targeted a sensitive area for Europe – its agriculture, on the other, it was seen as an opportunity to allow Russian agro-business to strengthen its position. After two years of such a protectionist regime, such policy seemed to be working out, with the exception of the variety in choice of products and higher food prices for Russian consumers:

> A handful of agricultural firms in Russia have stated that they fear sanction relief will undercut their expansion efforts. But while Russia cannot make brie cheese or grow certain fruits, it can pull more cod from the sea and produce more chicken from local farms. These firms and the Russian government that backs them with subsidies, are to some degree happy without Europe. As for the government, it retains rubles in the country. For food importers, it does away with exchange risk. Consumers, on the other hand, have had to face the brunt of the sanction regime. Import substitution is a long process and so prices have gone up.
>
> (Rapoza, 2016)

In March 2014, Russian President Vladimir Putin proclaimed that Russian consumers and the Russian market should be protected from GMO products. He also noted that work in this area should be carried out 'carefully' so not to violate the obligations of the Russian Federation in the WTO (Rosbalt, 2014).

At approximately the same time, Medvedev announced Russia 'does not need GMOs' and would be better off with organic agricultural products, mentioning the possible health risks of GM food for Russian consumers. He also argued that the Russian nation should be able to feed itself and do so without genetically modified products. He then noted that these restrictions should not affect scientific research on GMOs carried out in accredited laboratories (Prime, 2014).

In 2015, the Government Commission responsible for legal amendments approved a draft law prohibiting the cultivation and breeding of GMOs in the Russian territory. New changes were made in the regulation of GM crop production in Russia. The proposed amendments to the law on 'state regulation in the field of genetic engineering activity', 'Seed', 'Environmental Protection' and the Code of Administrative Offences introduced a legal responsibility for the unsanctioned production of GM crops (Vedomosti, 2015).

Russian policymakers have also confirmed 'the right of the Russian Federation' to ban the import of GM crops, due to the case that negative effects from GM crops on human health and the environment could be proven (Zamahina, 2015). Interestingly, Russian policymakers do not have a unilateral position on GM crops. Certain influential officials are known for their strong opinions. For example, Vice Prime Minister Arkady Dvorkovich is a well-known opponent to the use of GM crops in agriculture, while the former Minister for Health, Gennady Onishchenko, and members of the Communist party are proponents. For example, Communist MPs voted against the new law, noting that non-GM products could not compete with their GM counterparts in terms of crop yields and benefits for farmers. They stated that the production of transgenic agriculture is 10 times more profitable (Zamahina, 2015).

The Russian Ministry for Science and Education also stepped into the debate to confirm the permit to conduct GMO research for scientific purposes. It was responsible for the formation of a network of laboratories to conduct testing for the identification of GMOs and to develop further techniques for testing (Zamahina, 2015). MPs from the country's largest party, United Russia, suggested a reduction in the allowed levels of GM residue from 0.9% to zero and, allied with organic farmers who made claims in support of such a proposal, argued that Russia should be independent from outside influence in its protection of the interests of its citizens (Lyalyakina, 2014).

The Russian authority that controls the quality of agricultural production, known as Rosselkhoznadzor, has used the GMO rhetoric to attempt to ban the export of animal feed from the EC on the grounds that soya-based animal feed contains GMO. In the pre-WTO period, Russia had lists of approved exporting companies. After the WTO entry, these lists were withdrawn and the new rhetoric allowed the Russian authorities to return to a more selective policy on agricultural exports (Gluhodedov, 2016). In 2015, Rosselkhoznadzor temporarily introduced a ban on the supply of products (milk and animal feed) from two European companies – Schils B.V. from the Netherlands and Josera GmbH & Co KG from Germany (Alimov, 2015).

Such protectionist policy cost the EU €5.3 billion [$5.9 billion], which was wiped off exports; according to the European Commissioner for Agriculture, Phil Hogan, €50 billion annually is spent by Brussels to counteract the cost of Russia's restrictions on European farm produce (Hirst, 2016).

On the contrary, opponents of GM crops praised 'the world's largest nation's ban' 'against Monsanto and the US-led GMO cartel' which 'concentrated minds on the essentials of life' (Engdahl, 2015) and also picked up the rhetoric of restored Russian patriotism: Russians I spoke with during a visit August 2016 to the Rostov region told me they realised that the taste of Russian food, such as tomatoes, was far superior to that of imported food, which is often artificially coloured and treated with chemical preservatives in order to extend its shelf life by looking fresh. Following the tumultuous collapse of the Soviet Union in the early 1990s, the corrupt Yeltsin government opened the doors for Western agribusiness giants like Kraft, Nestlé, and Unilever to fill Russian stores with their agribusiness industrialised food products (Engdahl, 2015).

In 2015, a leading Russian channel showed a film about GM crops and global politics entitled 'Seeds and Darnels' made by two Russians, Konstantin Semin and Nikolay Diakov. The film is framed within the political context of the Ukrainian conflict. The film-makers went to Ukraine and spoke to the Minister of Agriculture and concluded that a new Ukrainian government formed from the opposition movement, Maidan, would lift the ban on GM crops and sell land to American corporations. It was suggested that 'Monsanto would enter into Europe from a back door – from Ukraine'. They also interviewed Vandana Shiva, who explained to Russian viewers that to destroy a farmer, to destroy a state is the recipe for monopoly', leading to market dependency and the suicide of farmers (Semin, 2015).

After the film had its national debut, I received phone calls from relatives who, like most Russians, retain small allotments and grow vegetables, asking me to promise not to grow GM crops. But, despite this newly introduced suspicion, most Russians do not participate in the GM debate, although there are a few civil society groups that do.

After a quick browse of the internet I found a few Russian NGOs discussing GM crops on their websites. I sent an email to one by the name of NAGS but did not receive a response. Then, in May 2015, at the closing ceremony at the GMO-free Europe conference, I found myself standing next to a young woman from NAGS, and we started a conversation which then led to my brief interview with the organisation in Moscow in October 2015.

NAGS (National Association for Genetic Safety) is a Russian NGO or, as it calls itself, a 'non-profit organization', the main activity of which is 'to 'contribute to the protection of biological and genetic safety of humankind and the environment, and to promote sustainable development ideas in human consciousness' (NAGS, date unknown). The NAGS brochure states that the organisation 'has had a great influence on the development of public debate concerning the safety of modern biotechnology, including GMOs, in Russia' (NAGS, date unknown). It lists six areas of activity: GMOs, sustainable development, food safety, organic agriculture, voluntary certification and genetic diversity conservation.

NAGS was founded in 2004 by Alexander Baranov (1946–2015) and Elena Sharoykina. Baranov was a Senior Research Associate at the N.K. Kolzov Institute of Biological Development who conducted his research in the area of studies on plant population and genetic diversity. He published three monographs and 70 articles. During 2000–2003 he was a member of a committee established by the Ministry of Natural Resources to provide guidance to the Russian government in regard to the management of GMOs. He was named as one of the experts who initially helped to block the commercialisation of GMOs. He also acted as a councillor to the Ministry of Energy, providing advice on technological regulations, and was a Russian representative to the International Commission on the Future of Food and Agriculture (Nauka i Zizn, 2007), a civil society network that brings together international activists, including Dr Vandana Shiva.

Elena Sharoykina was born in Ukraine in 1979 and graduated from the Dnepropetrovk National University with a Masters in Journalism. In the 1990s, she worked as a journalist in Ukraine covering political affairs and founded the Dneprovsky Centre in Social Studies (Russia.ru, 2012). She moved to Moscow to run her own consulting and media business and worked as a TV presenter at one of the leading Russian TV channels – Rossia – from 2008–2009. NAGS was formed as a result of her interest in environmentalism, and it is funded with profits from her media firm. This was her conscious choice, since it allows the organisation 'to be independent and not [to] take grants', although it is accepted that crowd-funding might be a funding option too (Sharoykina, 2015). In 2015, according to its brochure, NAGS had six members; two co-founders, a PR director, creative director, project manager and fundraising coordinator (NAGS, date unknown).

Elena is extremely well-established in Russia. For her TV series entitled 'Bio-razvedka', which translates as bio-intelligence, she interviewed Russian celebrities and European scientists about their views on GM crops. She also made contact with Skolkovo and Russian politicians including Vladislav Surkov, the former head of the president's office (Sharoykina, 2015). During our interview, I met two established officials from the Russian State Duma, one responsible for 'agrarian questions' and another for risk analysis and cooperation within the WTO and other international organisations.

To my question as to how many NGOs in Russia work on the question of GM crop debates, they answered that there are less than 10. One is a network called Hraniteli (translated from Russian as The Guardians), which defines itself as 'like-minded people interested in [the] preservation and development of ecological agriculture, [and] consumers rights', supporting the idea of local production and tasty, pure, honest food (Hraniteli, 2016). It is also a member of the Slow Food International whose programme Slow Food Kovcheg runs projects to preserve national agricultural biodiversity, promote healthy eating, raise awareness of the cultural and gastronomic traditions of the nations of the world and encourage food education and youth education projects. Another partner is Pravda o ede (translates as 'Truth about Food'). Both Hraniteli and Pravda o ede are managed by the same circle of people and are small organisations. International NGOs such as World Wildlife Fund (WWF) and Greenpeace do not participate in the Russian public debate on GM. Indeed, Greenpeace has even closed down its programme on GM crops in Russia, as Elena pointed out. Sharoykina's colleagues told me that President Putin is against GMO and that two areas in Russia, Krasnodar and the Far East region, are the main producers of organic soya. Further, it was pointed out that even American producers are aware of this fact. An interesting section of the website is on bioterrorism. As defined by NAGS, bioterrorism is the use of biological agents and toxins to destroy humans, food (including agricultural products), biological and ecological resources or ensuring control over these resources. It classifies the following types of bioterrorism: economic, ecological, food, genetic and agro-terrorism:

> In Russia in recent years, a number of factors have made bioterrorism real. On the one hand a worrying geopolitical situation, massive migration and the lack of adequate border control. On the other, there is the lack of a technical base combined with poor materials for revisionary authorities, the decline of the national health system, corruption, and the lack of responsibility and competence among officials. Under these circumstances, it is necessary to inform members of the authorities and citizens through the media and educational institutions, create resources for prevention, timely identification and responses to the results of bio-terrorist attacks.
>
> (NAGS, 2004)

On her TV programme, Sharoykina mentions the global conspiracy theory, which hypothesises that 'the golden million of the richest people in the world' aim to get rid of the majority of the population with 'the introduction of GM food which would eventually end human survival' (Russia.ru, 2009).

In one of its first media briefings, NAGS held a discussion in which Olga Razbash took part. She is a member of the previously mentioned NAGS partner, Hraniteli, and the Russian Regional Ecological Centre, another NGO relatively active in the debate. Before the briefing, Razbash took part in a meeting with NATO experts in Liege in May 2004 and concluded that after the 9/11 and Madrid attacks, the next target for terrorism would be food (NAGS, 2004).

Similar to Danube Soya, NAGS runs its own voluntary certification system called Biologicheski Bezopasno (which translates as 'Safe in biological terms'). This voluntary labelling system is officially registered with the Russian state authorities and works with accredited laboratories which run tests. After testing, producers receive the right to use the label on packaging and convince consumers that the product is safe for human consumption. Throughout the validity period of certification, NAGS runs quality control over the certified products of its member companies. Members include Russian companies such as 'Mihalichi' (a family-run farm), an agro-firm called Katusha, Zarizino which is an eco project and Russequelle, a producer of mineral water (NAGS, unknown date).

An annual campaign organised by NAGS since 2012 serves as a protest against Monsanto and has followed the agenda of the Occupy Monsanto social movement (KP, 2015). In 2013, NAGS organised a petition entitled 'No Monsanto. Russia without GMO!', aimed at the collection of one million signatures to an open letter to President Putin (NAGS, 2013). In 2014, a flash mob called 'Ludi I Zveri vs GMO' (which translates as People and Animals against GMOs) was organised by NAGS and its partner NGO Pravda o ede. In 2015, the same alliance ran a meeting to support GM food-labelling. This time the venue was outside the US Embassy in Moscow. One activist, dressed as sweetcorn, tried to hand in the petition against GMO to the Embassy but was stopped at the gate (Gluten-free, 2015).

Danube Soya and the National Association for Genetic Safety are both examples of how NGOs fit into regional politics. Regionalism brings a new political understanding of the concept of food security. It refers to state capacity to provide enough safe food for citizens while supporting and protecting national agricultural producers from external competition and instability imposed by other global market players, particularly international agribusiness. An understanding of individual regions might also be amended to accommodate new political interests. Protectionist policy is still a reserved option for states such as Germany and Russia that have agricultural resources but are not the top international competitors. As both states are members of WTO, and Germany is a member of the EU, neither can fully implement protectionism, thus anti-GM rhetoric becomes a useful tool to excuse some protectionism. This also explains why both states support NGOs promoting anti-GM discourse and regional GM-free zones.

Literature

Ackrill, R. (2000) *The common agricultural policy.* Contemporary European Studies, 9. Sheffield: Sheffield Academic Press.

Alimov, T. (2015) Rosselhoznadzor zapretil vvoz tovarov predpriatiy iz EC iz-za GMO, *Rossiaskaya Gazeta*, 25 September 2015. https://rg.ru/2015/09/25/ogranichenie-site-anons.html as viewed 01.07.2016.

ARC (2014) Food sovereignty: An idea whose time has come, Posted 16 June 2014. www.arc2020.eu/2014/06/food-sovereignty-movement-gaining-momentum as viewed 24.05.2016.

ARC (2016) Who we are. www.arc2020.eu/who-we-are as viewed 24.05.2016.

Behance (2014) Danube Soya. www.behance.net/gallery/14675085/Danube-Soya as viewed 27.05.2016.

Castells, M. (2000) *The rise of the network society.* Malden: Blackwell Publishing.

Castells, M. (2004) *The power of identity.* Malden: Blackwell Publishing.

Danube Soya (2015) Danube Soya. www.donausoja.org as viewed 01.07.2016.

Danube Soya (2016) Danube Soya Session & B2B Meeting in Vienna 21 April 2016. Danube Soya, Vienna.

De Visser, C.L.M., Schreuder, R., Stoddard, F. (2014) The EU's dependency on soya bean import for the animal feed industry and potential for EU produced alternatives. OCL, 21(4), open access.

Engdahl, W. (2015) Victory! World's largest nation bans GMO food crops. *NEO*, 1 October 2015. http://journal-neo.org/2015/10/01/victory-worlds-largest-nation-bans-gmo-food-crops/ as viewed 01.07.2016.
Europabio (2012) Bureaucratic barriers to biotech. Video posted 02.07.2012. www.europabio.org/bureaucratic-barriers-biotech as viewed 27.05.2016.
FAO (1996) *World Food Summit: Plan of Action*. Rome: FAO. URL: www.fao.org/docrep/003/w3613e/w3613e00.htm as viewed 2.05.2017.
Gaugitsch, H. (2015) Growing availability of Sustainable and GMO-free soya. Speech at Danube Soya East West protein forum. Berlin, 7 May 2015.
GENET (2003) Conference 2003: GE free zones in Europe conference: Examples, experiences, strategies. www.gmo-free-regions.org/past-gmo-free-conferences/gmo-free-conference-2003.html as viewed 01.07.2016.
GENET (2016) GENET-European NGO network on genetic engineering. www.genet-info.org/ as viewed 27.05.2016.
Gluhodedov, A. (2016) Rosselhoznadzor mozet zapretit vvoz kormov iz Evropy iz-za GMO, Izvestia, 23 May 2016.
Gluten-free (2015) V Moskve sostoyalsa miting v podderzky markirovki produktov s GMO. www.glutenlife.ru/news/10455.html as viewed 01.07.2016.
GMO-free Europe (2005a) Berlin manifesto for GMO-free regions and biodiversity in Europe. www.gmo-free-regions.org/past-gmo-free-conferences/gmo-free-conference-2005/berlin-manifesto.html as viewed 01.07.2016.
GMO-free Europe (2005b) List of participants. www.gmo-free-regions.org/past-gmo-free-conferences/gmo-free-conference-2005/list-of-participants.html as viewed 01.07.2016.
GMO-free Europe (2015a) Berlin declaration. Berlin, 8 May 2015.
GMO-free Europe (2015b) Danube Soya association East West protein forum. www.gmo-free-europe.org/program-info/7th-of-may-2015/danube-soya-association.html as viewed 27.05.2016.
Gordillo, G., Mendez Jeronimo, O. (2013) *Food security and food sovereignty (Base document for Discussion)*. Rome: FAO. www.fao.org/3/a-ax736e.pdf
Hausling, M. (2011) *Report: The EU protein deficit: What solution for a long standing problem*. Strasbourg: European Parliament. 2010/2111(INI).
Hirst, M. (2016) EU farmers suffer up to $6 bln in export losses from Anti-Russia sanctions. *Sputnik*, 20 May 2016. http://sputniknews.com/europe/2010520/1039956598/eu-sanctions-farmers.html as viewed 01.07.2016.
Hraniteli (2016) Missiya. http://anohraniteli.ru/about-us/mission/ as viewed 01.07.2016.
Kollmorgen, W.M. (1945) Crucial deficiencies of regionalism, the American economic review. *Papers and Proceedings of the Fifty-Seventh Annual Meeting of the American Economic Association*, 35(2): 377–389.
KP (2015) Piket proshel v ramkah vsemirnoy nedeli deystviy protiv geneticheski modifizirovannyh organismov. 23 September 2015. www.kp.ru/online/news/1254275 as viewed 01.07.2016.
Krön, M. (2015a) Opening speech: Danube Soya East West protein forum. Berlin, 7 May 2015.
Krön, M. (2015b) Speech: Evening reception in the Austrian embassy. Berlin, 7 May 2015.
La Via Campesina (1996) The Right to Produce and Access Land. http://www.acordinternational.org/silo/files/decfoodsov1996.pdf as viewed 02.05.2017.
La Via Campesina (2011) Organisation: The international peasant's voice. www.viacampesina.org/en/index.php/organisation-mainmenu-44 As viewed 25.05.2016.
Lee, R. (2007) Food security and food sovereignty. Discussion Paper N11. Centre for Rural Economy: University of Newcastle upon Tyne. March 2007.

Lyalyakina, A. (2014) Deputaty predlagaut zapretit proizvodstvo produktov s GMO, Izvestia, February 3 2014. http://izvestia.ru/news/564969 as viewed 01.07.2016.

Medvedev, D. (2010) Ukaz Presidenta RF ot 30 Yanvarya 2010. N 120 Ob utverzdenii Doktriny Prodovolsvennoy Bezopasnosti. http://base.garant.ru/12172719/ as viewed 01.07.2016.

Meunier, S. (2005) *Trading voices: The European union in international commercial negotiations*. Princeton: Princeton University Press.

NAGS (2004) Briefing OAGB Genetichesky I prodovolsvenny terrorism. 22 June 2004. www.oagb.ru/bio.php?txt_id=131 as viewed 01.07.2016.

NAGS (2013) Novosti. Akzii protiv GMO proshli po vsemu miry. 27 September 2013. www.oagb.ru/info.php?txt_id=17&nid=14411&page=0 as viewed 01.07.2016.

NAGS (Date Unknown) *All-national association of genetic safety*. Moscow, Russia.

NAGS (Date Unknown) Biologicheski Bezopasno. Sistema kontrolya kachestva produktov. Moscow, Russia.

Nauka I Zisn (2007). Geneticheski Modifizirovannie Organizmy. Interview 17 December 2007, Moscow. www.nkj.ru/interview/11630/ as viewed 01.07.2016.

Nye, J.S. (1968) *International regionalism readings*. Boston: Little, Brown and Company.

Pelagidis, T., Papasotiriou, H. (2002) Globalization or regionalism/states, markets, and the structure of international trade. *Review of International Studies*, 28: 519–535.

Prime (2014) Medvedev: Rossia ne budet eksportirovat produkty s GMO. www.1prime.ru/News/20140405/782030840.html as viewed 01.07.2016.

Raikes, P. (1988) *Modernising hunger: Famine food surplus and farm policy in the EEC*. London: Catholic Institute for International Relations.

Rapoza, K. (2016) Nearly two years into sanctions, can Russia live without Europe? *Forbes*, 19 April 2016. www.forbes.com/sites/kenrapoza/2016/04/19/nearly-two-years-into-sanctions-can-russia-live-without-europe/#730bd497211e

Rosbalt (2014) Putin: Nuzno zapretit GMO, sobludaya pravila WTO. www.rosbalt.ru/main/2014/03/27/1249489.html as viewed 01.07.2016.

Russian Federation (2010) Ukaz ob Utverzdenii 'Doktriny prodovolstvenoy bezopastnosti Rossiskoy Federazii' February 1 2010 http://kremlin.ru/events/president/news/6752

Russia.ru (2009) 3 mlrd chelovek mogut umeret. www.russia.ru/video/bio_8618/ as viewed 01.07.2016.

Russia.ru (2012) Elena Sharoykina. Dossie. www.russia.ru/hero/sharojkina/profile/ as viewed 01.07.2016.

Semin, K. (2015) Semena I Plevela: genno-modifizirovanny sled Zapada na Maidane. *Vesti*, 24 March 2015. www.vesti.ru/doc.html?id=2450749 as viewed 01.07.2016.

Sharoykina, E. (2015) Interview with Elena Sharoykina. Moscow, Russia, 22 October 2015.

Swinbank, A., Tanner, C. (1996) *Farm policy and trade conflict: The Uruguay round and CAP reform*. Ann Arbor: University of Michigan.

Torgovo-Promyshlennaya Palata (2012) Tema: Selskoe Hozyastvo Rossii: Vozmozen li trehkratny rost? 22 November 2012. http://me-forum/media/events/mef-founding2 as viewed 01.07.2016.

Ushachev, I. (2008) *Osnovny Polozenia Dokrtiny Prodovolstvennoy Bezopasnosti Rossiskoy Federazii. Ekonomika*. Upravlenie: APK.

Vayrynen, R. (2003) Regionalism: Old and new. *International Studies Review*, 5: 25–51.

Vedomosti (2015) Rossija polnost'ju prekratit proizvodstvo produktov s GMO. 18 September 2015. www.vedomosti.ru/politics/news/2015/09/18/609332-rossiya-prekratit-proizvodstvo-gmo as viewed 01.07.2015.

von Cramon-Taubadel, S. (2000) The reform of the CAP from a German perspective, in W.P. Grant, J.T.S. Keeler (eds.) *Agricultural Policy*, Vol. 1. Cheltenham: Edward Elgar, pp. 411–426.

Wilson, G.A., Wilson, J.O. (2001) *German agriculture in transition: Society, policies and environment in a changing Europe*. Basingstoke: Palgrave.

WTO (2012) Accessions Russian federation. www.wto.org/english/thewto_e/acc_e/a1_russie_e.htm as viewed 01.07.2016.

Zamahina, T. (2015) Deputaty zapretili ispol'zovat' GMO v agropromyshlennom komplekse. *Rossiskaya Gazeta*, 24 April 2015. https://rg.ru/2015/04/24/gmo-site-anons.html

5   The sustainability discourse

The conceptual framework of sustainable development[1]

Sustainability is arguably the most important conceptual framework in modern policymaking. The classic conceptualisation of sustainable development can be found in the Brundtland report. Although originally the concept of sustainability came from agricultural studies as it was offered by Wes Jackson (Jackson, 1985). The term was picked up and further developed and institutionalised under the support of the United Nations. The World Commission on Environment and Development (WCED), chaired by the Norwegian former doctor and politician Gro Brundtland, published its report 'Our Common Future' in which the concept of sustainability was formulated in its classic form in 1987. It recognised that there was a need for 'a global agenda for change' at that time (WCED, 1991, p. ix).

The epoch, the 1980s, when the report was written, saw a number of crises, economic and ecological ones, and raised serious questions about changes in management of environmental resources and addressing social and economic issues. Chernobyl catastrophe and Bhopal disaster were serious reminders about the importance of the precautionary principle and the need for better management of environmental resources. So the main global challenge to sustainability was to build up a strategy of socio-economic development for fast growing populations under the condition of a decaying environment and finite natural resources. The publication and its further acceptance and support meant that for the first time, at the international level at least, there was an understanding that environmental problems, social development and economic growth were highly interlinked. At policy level, this meant that decisions concerning economics should address both environmental and social concerns.

According to the report, sustainable meant the ability '[to meet] the needs of the present without compromising the ability of future generations to meet their own needs' (WCED, 1991, p. 8). In other words, development should have been accompanied with long-term assessment and planning. The critical objectives for policies under the sustainability framework were identified as reviving and changing the quality of economic growth; meeting the basic needs of the population for jobs, food, water, energy and sanitation; environmental conservation; ensuring a sustainable level of population; and reorienting technology to address these issues.

The authors of the report were realistic about the implementation of a new model of development, understanding that

> sustainable development [was] not a fixed state of harmony, but rather a process of change in which the exploitation of resources, the direction of investments, the orientation of technological development, and institutional change are made consistent with future as well as present needs.
>
> (WCED, 1991, p. 9)

They predicted that there would be 'painful choices to be made' (WCED, 1991, p. 9). Economic growth always brings risk of environmental damage, as it puts increased pressure on environmental resources. Thus, the main question of sustainable development is how to ensure sustainable (meaning long term), economic growth with sustainable (meaning responsible) resource management and accounting for social development.

The question about ensuring economic growth and environmental conservation, left open by the report, has produced discussion between the economists who argued for weak sustainability (focus on economics) and those who argued for strong sustainability (focus on ecological conservation) (Pearce and Atkinson, 1998). The Brundtland commissioners did not have a ready-made solution for this dilemma but recognised that each country would have to find its way and hoped that international cooperation could smooth the path.

However, at least one possible tool to achieve sustainability was named in the report – the reorientation of technologies to provide a key link between humans and nature and development of alternative technologies which could help 'to produce more with less'. Yet it was also recognised that environmental risks could also arise from these technological advances – the commissioners remembering the tragic examples of environmental catastrophes such as Chernobyl. So on the one hand, the commissioners warned against 'a blind faith in science's ability to find solutions', but on the other, they still believed that technology was a vital tool for achieving sustainability (WCED, 1991, p. 102).

The report set the target for achieving sustainable development as the year 2000. However, the target has not been met, and this has raised serious debates about its achievability. Recent authors, such as Kos (2012), even have asked if sustainable development has been an altogether utopian idea.

The report introduced two more specific sub-issues that have been central to the GM crops discussion, so it is worth remembering how they were understood back in the 1980s. These are food security and biodiversity.

The Brundtland report articulated the specific issue of biodiversity termed in the report as biological diversity. In its discussion, the commissioners raised concerns about the extinction of species and deterioration of ecosystems such as coral reefs and referred to 'genetic variability' which is used in the utilitarian sense: genetic diversity was seen as vital for the survival of species and enhancing genetic engineering by humans. The report predicted 'Gene Revolution' which would allow 'harvesting crops from deserts, from seawater, and from other environments that did not previously support farming' (WCED, 1991, p. 156). Thus, the most varied genetic pool, the 'genetic wealth', was aimed to serve both – conservation of species and ecosystems, and provide grounds for technological advances that could promote economic growth. To compare, the recent definition of biodiversity offered by UNEP (2010) is very close to the WCED's: 'Biodiversity is the variety of life on Earth, it includes all organisms, species, and populations; the genetic variation among these; and their complex assemblages of communities and ecosystems' (UNEP, 2010, p. 2). The following activities threaten biodiversity: destruction of habitats, alterations in the composition of ecosystems, invasion of alien species, overexploitation, pollution and contamination, and climate change (UNEP, 2010, p. 3). Generally, there is a consent that biodiversity is valuable and should be protected.

Food security is a specific issue that has been much discussed by the Brundtland report and its followers. The World Food Summit in 1996 defined food security very closely to the terms of sustainability, as 'physical and economic access, at all times, to sufficient, safe and nutritious food (for people) to meet their dietary needs and food preferences for an active and healthy life' (cited in Swaminathan, 2003).

The main questions discussed were demands of growing populations, which meant increase in productivity of and improved access to food. By the time of the Brundtland report, agriculture had already achieved greater increase in productivity of food per head 'than ever before in human history' (WCED, 1991, p. 118). However, there were a number of signs of crisis in agriculture and serious challenges ahead, including the deterioration of natural resources, climate change and increasing demand for food, so the priority of increasing productivity in agriculture remained on the agenda of the report. With the introduction of the sustainability approach, a new focus had been added: more food should be produced with fewer resources. The possible solutions offered by the report included reforming government interventions, improving land and water management, finding alternatives to chemicals, strengthening the technological and human resource base for agriculture, promoting equity in food production and distribution and rural development (WCED, 1991, pp. 130–143). Environmental protection was linked to economic growth and social welfare.

Sustainability, as one of the frameworks, has been used in the GM debates. It is often used by international organisations and business. For example, after her chairing of WCED, Brundtland moved to the World Health Organization (WHO) and remained as the senior international politician with a respected voice. In regards to GM debates, she was involved in the discussion of the GM food as food aid in Southern Africa. In 2002, she issued a joint statement with the UN Secretary-General's Special Envoy for the Humanitarian Crisis in Southern Africa Mr Morris, where they stated that food donations received for Southern Africa in 2002 contained GM maize and that according to FAO and WHO, this was certified food 'safe for human consumption' (WFP, 2002). In a related meeting in Harere, Ethiopia, in August 2002 with health ministers from 10 Southern African countries, Brundtland said that GM foods are 'not likely to present human health risk', and that WHO 'is not aware of scientifically documented cases in which the consumption of these foods has had negative human health effects' (WHO, 2002).

Among agribusinesses, sustainability is often regularly used as a framework to promote GM crops. For example, Monsanto adopted a slogan in 2008 'The Challenge: Meeting the needs of today while preserving the planet for tomorrow' and accepted 'a commitment to sustainable agriculture' (Bjerg, 2016).

NGOs often refer to sustainability terminology and, in fact, most of the smaller debates by NGOs fit into a three-dimensional model of sustainable development: there are debates about environment and human health, about economic efficiency and about social impact of GM crops. Briefly, one can also discuss how a recently argued fourth pillar of the sustainable development, which is culture, is used in the debate.

Arguments about effects of GM crops on the environment and human health

The main issue in these debates is centred on the safety of GM plants to human health and the environment. The possible risks from GM crops can be summarised as 'the risk to human and animal health through eating genetically modified food', 'adverse or unforeseen events caused either by mobility of the transferred gene from the new GM variety into other species', and 'deleterious environmental outcomes on the arable ecosystem and its associated wildlife' (Perry, 2003, p. 43).

The debate on the possible human health risks from GM crops started with the scientific community from the Pusztai research controversy. Dr Árpád Pusztai, working on a GM potato project in the Rowett Institute in the mid-1990s, concluded that GM potatoes had negative effects on rats. After his appearance on a TV programme and a further publication in *The Lancet* in 1999, the whole debate about the safety of GM foods was activated. The Pusztai affair is discussed in more detail in Chapter 8.

Since then, the potential risks of GM crops to human health have been a constant argument used by the oppositional NGOs. The opponents of GM crops, such as the Earth Open Source organisation, claim that there are serious health risks leading to damage in most systems of human body: 'Most studies with GM foods indicate that they may cause hepatic, pancreatic, renal and reproductive effects and may alter haematological, biochemical and immunological parameters, the significance of which remains to be solve with chronic toxicity studies' (Antoniou et al., 2012). Earth Open Source is an NGO set through private donations, including the Isvara Foundation, 'to restore the open source roots of the food system – collaboration, transparency, and shared knowledge and resources – to help feed humanity, increase equity, support self-reliance and foster healthy ecosystems'. It also runs its own Earth Open Source Institute. In 2017, their website names John Fagan as its executive director (Earth Open Source, 2017).

Before the current restaffing, there were two natural scientists Michael Antoniou and John Fagan and one social scientist Claire Robinson associated with the NGO, and they appeared as authors of the reports published by Earth Open

Source. Dr Michael Antoniou studied biochemistry in Oxford and currently runs his own research group in Kings College, London. His Gene Expression and Therapy Group (GETG) has

> the main research focus of the GETG is the characterisation of genetic regulatory elements with a dominant chromatin remodelling (opening) capability and their exploitation in the development of efficient expression vectors for efficacious and safe biotechnological applications namely protein therapeutic biomanufacturing and human gene therapy.
>
> (Antoniou, 2016)

Dr John Fagan received his education from the University of Washington and Cornell University. According to his LinkedIn page, he worked on the boards of the two GMO testing group companies Genetic ID and Global ID. In 2012–2014, he worked as director for Proterra Foundation. He is also 'professor of molecular biology at Maharishi University', 'where he leads a research program using transcriptomic, proteomic and metabolomics approaches to understand the effects of Transcendental Meditation (TM) on gene expression (Fagan, 2016). In 1994, he 'made a few headlines' returning a grant from the National Institutes of Health to conduct research in Fairfield, Iowa, and turned to social activism against GMO. In 1996, prior to Pusztai, he called transgenic food 'a dangerous global experiment in which we are the guinea pigs' (Copple, 2000, para. 2).

The NGOs who are supporters of GM crops, such as the International Service for the Acquisition of Agri-Biotech Applications (ISAAA), argue the contrary.

ISAAA is a 'not-for-profit international organization that shares the benefits of crop biotechnology to various stakeholders, particularly resource-poor farmers in developing countries, through knowledge sharing initiatives and the transfer and delivery of proprietary biotechnology applications' (ISAAA, 2016). According to LobbyWatch (2016), its funders include big agribusinesses such as Bayer, Monsanto and Syngenta and have their own three research centres.

In the GM debate on health safety, ISAAA states that GM foods pass food safety assessments, which include such as toxicity, allergenicity and antibiotic resistance, and refer to the studies by international organisations showing that GM foods are safe. They refer to the international agencies such as the Food and Agriculture Organization (FAO), World Health Organization (WHO), European Commission (EC), Académie Nationale de Médecine (French Academy of Medicine, ANM), American Medical Association (AMA) and the USA Society of Toxicology (SOT) who have reviewed these health issues and have come to an agreement that GM foods are safe for human health' (ISAAA, 2009).

The same argument was put by Prof. Ammann, another biologist involved in the GM debate and activism, its pro-GM side. He stated,

> None of the predictions about negative health effects have materialised. This is why many [campaign groups] have now changed their strategy and are now more careful with allegations about alleged adverse health effects. They have

switched instead to occasional anecdotal evidence as well as raising doubts about 'long term effects.

(EuropaBio, 2013, p. 18)

However, the opposition in response to such statements makes claims that long-term risks may show themselves later and, until then, long-term continuous study is needed. This was precisely the argument used by Dr Antoniou at a debate in Belfast where he urged for more long-term monitoring of potential risks (Antoniou, 2014).

Oxfam's research, however, has acknowledged that the reduced use of pesticides due to the new transgenic seeds has shown a reduction in accidental pesticide poisoning (Tripp, 2009). It is also remarkable that both the reports from opponents and supporters, which have been cited earlier, use the same form in presenting their arguments: both refer to 'myths about GM' and then compare them to 'reality' in order to persuade their readers.

The next issue debated is the impact of GM crops on the environment, and it has also produced a diversity of opinions in the scientific community[2] and civil society. The main types of GM crops (resistant to the herbicide Roundup Ready – glyphosate and containing an insecticide from the bacterium Bacillus thuringiensis – Bt) are designed to increase a plant's resistance to weeds and insect pests respectively. This has been interpreted differently by both opponents and supporters in order to prove their point of view.

Originally, the argument in support of using Bt came from the environmental movement itself. Rachel Carson, in her famous book about the environmental damage caused by human activities, wrote about the 'high hopes on Bacillus thuringiensis, found in Thuringia, Germany, that kills by poisoning larvae, can be used to stop crop damage'. For her it was an ecological alternative to DDT (Carson, 1963, p. 289). The next generation of environmentalists, who now oppose GM crops, raised their concerns about using the Bt technology. It was expected that pests could evolve resistance and that the insecticide would thereby be rendered ineffective.

Antoniou et al. expanded the same argument in their study, claiming that GM crops 'do not reduce pesticides use but increase it', 'create serious problems for farmers, including herbicide-tolerant "superweeds", compromised soil quality, and increased disease susceptibility in crops' and 'harm soil quality, disrupt ecosystems, and reduce biodiversity' (Antoniou et al., 2014, p. 18). While farmers have recorded the increasing resistance in weeds and pests using conventional pesticides and herbicides (Richardson, 2001), it is possible to argue that the building up of resistance is a fundamental issue in biology. It does not mean though that all pest treatments should be given up or that new solutions will be found easily. Anti-microbial resistance presents a similar challenge in human and veterinary medicine. The introduction of antibiotics has reduced human deaths from inflammatory diseases from 43% to a mere 7%. However, without measures to address current growing anti-microbial resistance, the death rates in the future are estimated to be 10 million people per year globally (Davies, 2015). Further scientific research is crucial in both cases of increasing resistance.

Vandana Shiva has made strong claims which do not allow any space for GM crops: 'Genetic Contamination is Inevitable. Coexistence is not Possible' (Shiva, 2011, p. 21). Ammann, however, countered that the 'intrinsic values of the plant genome', which opponents fight to protect so hard is 'a fiction' because of evolutionary and breeding processes (Ammann, 2008, pp. 3–4).

In parallel, Ammann has also referred to the same concept of biodiversity but, in turn, he claimed that scientific studies do not show 'permanent negative impact on biodiversity done by genetically engineered crops'. On the contrary, 'agricultural biotechnology is a real help for maintaining biodiversity', as it reduces the use of pesticides, and by having greater yields from less land, would help environmental conservation (Ammann, 2006). He has also suggested learning best practices from organic farming such as techniques to control pests and weeds in an ecological way, to use traditional knowledge and to integrate them in what he has called 'precision farming' which would use GM seeds and grow them in an organic way (Ammann, 2008, 2009, 2014).

This idea of merging GM and organic agriculture, however, does not meet support from the opposition. Antoniou and his co-authors have counter-argued that GM crops do not reduce pesticide use but, on the contrary, increase dependency on agrochemicals. The consequent spread of 'glyphosate-resistant superweeds' makes farmers spray more (Antoniou et al., 2012, p. 74). Since the spray is at high level, it leads to negative environmental impacts on farmers, consumers and ecosystems. The use of no-till or low-till farming, which is beneficial to soils, is already practiced and 'farmers do not have to adopt GM crops or use herbicides to practice no-till' (Antoniou et al., 2012, p. 76).

This brief review of the arguments from the GM crop supporters and their opponents shows that both of them claim to favour preserving biodiversity and conserving nature, but there are fundamental differences in their views on the consequences of using biotechnology in agriculture.

Economic argument

The major issue discussed in the economic debates on GM crops are the yields, input costs and economic benefits to farmers and consumers when GM and organic agricultural techniques are compared.

The fact of the growth in commercial production of GM crops has been used as an argument in their favour. According to ISAAA, 'in 2014, the global area of biotech crops continued to increase for the 19th year at a sustained growth rate of 3 to 4% or 6.3 million hectares (~16 million acres), reaching 181.5 million hectares or 448 million acres' (James, 2014). ISAAA Briefing 49 reported,

> Latest global provisional information for the period 1996 to 2013 shows that biotech crops increased production valued at US$133 billion; in the period 1996 to 2012 pesticide use decreased significantly saving approximately 500 million kg of active ingredient. In 2013 alone, crop plantings

lowered carbon dioxide emissions equivalent to removing 12.4 million cars from the road for one year.

(ISAAA, 2014)

Dr Shiva accepted the fact that GM crops have been growing worldwide (Shiva, 2011, p. 14), but she confronted the data about the yields, at least in regards to the Bt cotton produced in India by the reference to the data she collected in India. According to her, Monsanto's claims that 15,000 kg of GM cotton produced per acre were not true, and the real average yields were only 400 kg. She did not believe that GM crops reduced the use of pesticides because of 'super weeds taking over your fields' (Shiva, 2011, pp. 10–11).

Other GMO opponents, such as Earth Open Source, denied the fact that it has been possible to engineer the yield, saying, 'High yield is a complex genetic trait based on multiple gene functions and cannot be genetically engineered into a crop' and referred to field tests which, in some cases as with Bt maize, suggested lower yields of GM crops in comparison with non-GM crops (Antoniou et al., 2014, p. 230). They then developed their argument about inefficiency of GM crops' yields by claiming that GM crops fail to yield more (Antoniou et al., 2014, p. 232). The same rhetoric is used to contradict the argument brought by the supporters of GM crops that these crops can be drought resistant (Lynas and Robinson, 2012). Robinson argued that 'high yields, disease resistance and drought tolerance are complex traits that are much easier to achieve with conventional breeding than GM' and that is why she believes that 'agro-ecology and conventional breeding' 'outperforms GM' (Lynas and Robinson, 2012).

ISAAA (2012) claimed that biotech crops have become the fastest adopted crop technology ever with 6% growth of cultivated hectarage, and the reason for that was 'it delivers benefits'. They reported that 'Bt cotton increased the income of farmers significantly by up to US$250 per hectare' (ISAAA, 2012, p. 2). According to their report, 'India enhanced farm income from Bt cotton by US$12.6 billion in the period 2002 to 2011 and US$3.2 billion in 2011 alone' (ISAAA, 2012, p. 2).

The reaction of the opponents of GM crops to these studies and the data produced about the economic benefits derived from GM crops was complete denial and raised doubts about the validity of the research (Antoniou et al., 2012, 2014) and accusations of 'deliberate ignorance' (Shiva, 2011).

Earth Open Source's members prefer to withdraw from the debates on the economic impacts of GM crops on farmers, claiming that the question 'is complex and a thorough examination is beyond the scope of this report' (Antoniou et al., 2014, p. 264). They accuse the studies which praise GM crops for creating economic benefits to farmers, such as Brookes and Barfoot (2004), for being 'not peer-reviewed' and dependent on the data from industry (Brookes and Barfoot, 2004, p. 266).

Dr Shiva has supported her argument about the lack of economic benefits for farmers from GM crops by making two points: one is the rise of the cost of seed.

She claimed that in India, the shift to Bt cotton meant a jump of 8,000% in the cost of seed. The other point was royalties that farmers had to pay to agro-companies owning the seed: for example, she stated that in India Monsanto was collecting Rs700 as royalty for a 450 gm packet of seed costing Rs1,600 (Shiva, 2013).

Moderate supporters of GM technology, such as associated researchers of Oxfam, have agreed that GM seeds have been more expensive than non-GM seeds (50–75% more costly than conventional seeds) (Tripp, 2009, p. 81). It is also stated that relatively high prices occur because of the prevalence of hybrids in the market. In certain cases, as of Bt cotton seeds prices in India, state governments have stepped in and threatened to ban the agro-companies from operating, which resulted in halving of the price of the seeds (Tripp, 2009, p. 78).

The suggestions of the GM supporters, such as Ammann (2009) and Lynas and Robinson (2012), who have argued that no-till GM agriculture can be beneficial in terms of climate change preparedness and saving energy costs as it avoids ploughing, also have been denied by the anti-GM activists. They raised concerns about Monsanto's application for carbon credits for GM Roundup Ready crops under the United Nations' Clean Development Mechanism, as it could lead to farmers' encouragement from governments to grow GM varieties. To prevent this, they referred to the studies showing little difference between carbon sequestering by no-till agricultural land and ploughed fields, and thus recommended organic agriculture (Antoniou et al., 2012, p. 102).

Those who support the application of GM crops have also compared new biotechnology-based agriculture with organic farming. They indicated that organic crops produced low yields because of higher level of exposure to fungi and pests and the fact that there were not enough natural fertilisers to put back into soils. Fungal diseases also meant contamination of crops with toxins which then required additional investment to address this issue (Heap, 2003). For Lynas (Lynas and Robinson, 2012) 'organic is not the answer', because farmers either 'use equally toxic copper sulphate, or simply watch the crops get devastated'.

In its turn, Greenpeace uses the critique of Green Revolution which took place in the global South and did not yield promised economic growth but rather damaged existing natural resources and argues that 'real green' would be global transfer into organic agriculture which would manage ecological resources and support farmers with good income (Parrot and Marsden, 2002).

GM crops and social development

The most discussed issues about GM crops' relevance to the social aspects of international development have been their impacts on small farmers in developing countries, possibility to alleviate hunger and malnutrition and capacity to provide food security.

Often those who support the use of GM crops refer to the fact of the world's fast growing population and the necessity to 'ensure food security for everyone' (Lynas and Robinson, 2012). They suggested that the high productivity of GM seeds and their lower ecological impact could have also been a possible solution to addressing the issue of hunger in the developing world. Africa, for example, as ISAAA hoped, could increase crop production from 1 billion hectares (15% of global) to 3.6 billion hectares (35% of global) and thus, in their opinion, address global food security (ISAAA, 2012, p. 3).

The views of humanitarian organisations, such as Oxfam, who work directly with these issues, are a good perspective from which to review this argument. Before GM crops were introduced into the agenda of hunger and poverty alleviation, Oxfam referred to the traditional knowledge of farmers and organic farming in their aid programmes in Africa, affected by drought and hunger in the early 1990s (Myers et al., 1992). And in the 1990s, Oxfam members, such as Dorothy Myers, considered organic cotton as a better alternative to GM cotton (Myers and Stolton, 1999).

Greenpeace also supported the idea of organic agriculture as an alternative to conventional agriculture using GM seeds and its social consequences. Its previously cited report 'The Real Green Revolution' criticised the Green Revolution for heavy use of pesticides and argued that organic agriculture which allows farming 'to jump off the chemicals treadmill' is the real green revolution, and that it will allow living standards of farmers to be improved in the developing world (Parrott and Marsden, 2002).

But the same argument that the lower use of chemicals is beneficial for farmers has been used in support of the use of GM crops. Oxfam's report has suggested that lower use of pesticides in cultivating transgenic cotton produced health improvements for people and saved time for other activities while organic farming has certain constraints.

Tripp, who edited the Oxfam report, suggested that the economic constraints of organic agriculture could also lead to negative social consequences. For example, organic low-input agriculture increases inputs of labour time, and the deficiencies of organic fertilisers do not enable enough crops to be produced to satisfy the growing food demand in Africa (Tripp, 2006, pp. 5–6).

In 2010, Oxfam published the report 'Biotechnology and Agricultural Development', which summarised the results from the Oxfam-America project entitled 'Learning from the Experience of Small-Scale Farmers: The Case of Transgenic Cotton'. It studied the experience of growing GM cotton in developing countries and discussed 'the relevance of agricultural biotechnology for resource-poor farmers' (Tripp, 2009). The authors admitted that they had 'to examine a very complex and controversial subject' and did not insist on the universality of their findings due to research methodology limitations, but the data they found proved the decrease in insecticides costs and increased yields resulting from adoption of Bt cotton occurred in a number of countries including China, India and South Africa (Tripp, 2009, p. 74). For farmers, it also meant fewer cases of pesticide-related health problems (Tripp, 2009, p. 76) and saving more time for farm management (Tripp, 2009, p. 81). The main factor which determined farmers' net returns was the balance between the costs of seeds and savings derived from the application of less chemicals and yield gains. The report considered Bt cotton as part of a broader strategy in increasing efficiency of agriculture and addressing farmers' well-being while insisting that it should be accompanied with other policy measures.

The environmental anti-GM NGO GRAIN criticised Oxfam-America for publishing the report as a book and denied the validity of its data. In their view, the report was a 'review of a very limited volume of existing data on the topic', which concentrated only on economic analysis of yields and profit, lacked neutrality and 'diverted attention from real solutions for smallholder and subsistence farmers: structural reform and ecologically based agriculture' (GRAIN, 2010). Another organisation, GMWatch, also expressed their 'concerns with the recent publication publicised by Oxfam America in support of agricultural biotechnology as a viable solution for addressing poverty faced by resource poor and subsistence farmers in developing country'. In their eyes, Oxfam appeared to be positioning itself as a 'good broker' for GM technology by 'false advertising on appearing neutral' (GMWatch, 2010).

A regional South African case study of the Makhatini Flats' farmers has become particularly much debated among the activists and researchers. The case study was researched and published by a number of academic groups, some suggesting that the need to spray less was one of the major reasons of local farmers choosing to switch to Bt cotton and that it reduced sickness in the local hospital (Merritt, 2003). The similar argument was made in the Oxfam report, as shown earlier (Tripp, 2009).

The Navdanya report showed 'a different story' about the Makhatini Flats, where the majority of farmers are in debt. The argument that the high price for GM seeds leads farmers to become indebted is also illustrated by the account of Indian farmers' indebtedness. The report argued that this indebtedness has led to over 250,000 farmers' suicides in India (Shiva, 2011).

Another much debated issue has been about patents on the varieties of GM crops, control of farmers over seeds and the role of corporate interests in promoting social injustice (Shiva, 2012b).The criticism that GM crops are an example of biased corporate interests has been also addressed by the supporters of GM crops. Lynas argued that 'publicly funded, non-commercial, non-patented applications of GM technology aimed at the reducing the use of toxic chemicals' should be supported (Lynas and Robinson, 2012). Robinson, however, rejected the idea, because 'public-private partnerships mean that while public money funds the research and development (R&D), the developed trait is sold to companies'. 'Only patents make it attractive', she argued (Lynas and Robinson, 2012).

The case of Golden Rice, a GM rice fortified with a precursor of Vitamin A, a lack of which causes blindness, developed with the support of the Gates Foundation and agro-corporations such as Syngenta and aimed at addressing the malnutrition of poor farming communities in developing countries as well as being developed commercially, has been considered by the opposition as a hoax that could lead the way to other GM crops (Shiva, 2001). In their turn, the supporters of GM crops, such as Patrick Moore, have developed a wide media campaign to change the opposition to Golden Rice, claiming that this confrontation and further delay in providing access to this GM crop means the loss of many children's lives in developing world and, because of this, opposition to Golden Rice is 'a crime against humanity' (Moore, 2014). GMWatch, in turn, argued that Moore's claims are 'bogus' and that he and Mark Lynas 'see Golden Rice as a useful weapon with which to attack the environmental movement, which seems to be their main platform for self-promotion' (GMWatch, 2013).

Culture and sustainability: another pillar?

The Brundtland report and associated speeches of its chair Gro Brundtland illustrate that, indeed, the concept of sustainability was centred on people and one may argue that it represents the concept of weak sustainability. Supporters of strong sustainability, particularly deep ecologists, criticised the report for not adequately acknowledging the real issues of ecology (Visvanathan, 1991). Brundtland did not see anything wrong in the anthropocentric approach of sustainability, where humans and Nature are seen as an interlinked system: "People influence the trends that destines the planet. The planet affects people" (Brundtland, 2015).

The only compromise that Gro Brundtland has adopted was to accept the forth element to the three pillar system – culture. Such approach means that the introduction of cultural relativism to the concept of sustainability would allow to explain different regional speed of implementation of sustainability and would also mean that the concept was not imposing a one size fits all approach for which the report has been previously criticised (Brundtland, 2013). At the UN level, the Commission for United Cities and Local Governments promotes 'culture as the forth pillar of sustainable development' through its Agenda 21 for culture (United Cities, 2011).

Generally, culture, as science, is another broad concept which receives a constructivist approach in postmodernity, that is, it is included in the discussion of perception and interpretation of institutions and political development and has divergent value systems (Giordano et al., 2010).

Some GM debate activists, particularly Vandana Shiva, have also applied the four pillar approach. For example, Shiva analysed the Ganges ecological crisis as 'the social, cultural, ecological and economic lifeline of India is under severe threat' (Salazar, 2010). Her major organisation Navdanya, which translates as 'Nine seeds', aims at 'the rejuvenation of indigenous knowledge and culture' and seed sharing and conservation. Seed conservation also includes the notion of culture: 'Conserving seed is conserving biodiversity, conserving knowledge of the seed and its utilization, conserving culture, conserving sustainability' (Navdanya, 2016). Culture and knowledge aspects also mean promotion of 'awareness on the hazards of genetic engineering, defended people's knowledge from biopiracy' and food rights (Navdanya, 2016). Seed exchange is more than just a mere exchange, it is 'the exchange of ideas, knowledge, of culture and heritage', an accumulation of 'traditions, of knowing how to work the seed'. Such knowledge is the basis of 'cultural, religious, gastronomic' and other social values and might provide solutions to address challenges of climate change (Shiva, 2012a).

The inclusion of culture the fourth pillar of sustainability allows Shiva to add more criticism to the model of weak sustainability. For example, she developed her concept of 'monoculture of mind'. Initially, monoculture was referred to the dominance of one crop in agricultural production. For example, in India, eucalyptus plantations for pulp industry were spread and even World Bank funded such pro-grammes as part of regional development and called it 'social forestry', meaning that included issues of social development with economic development. In this case, the funding aimed to enhance supply of minor forest products and improve institutional capacities of Indian forestry and, ultimately, better manage natural resources of Indian forests (WB, 1994). At the same time, the spread of eucalyptus trees destroys soils for other trees. And the Navdanya farm in the Dehradun region was an example of that when Shiva just bought it and started to renovate (Bhatt, 2015). To counter-argue with the World Bank study, Dr Shiva and her colleagues from the Indian Institute of Management in Bangalore produced their own report 'Ecological Audit of Eucalyptus', which challenged the idea of eucalyptus planta-tions as forests and discussed a 'monoculture of the mind' as 'a blindness to diver-sity and its potentials, a blindness that blocked out the productivity of biodiverse systems in forest, in agriculture, in the ocean' (Shiva, 2014, p. 2). The opposite of monoculture of mind is cultural diversity.

Seed sharing and conservation was used to remind farmers about 'forgotten foods' from their cultural heritage and reintroduce crops which were used by their ancestors, such as varieties of dal and millet and amaranth (Shiva, 2014). At the farm, these seeds are kept in the bank and dried samples of plants are used to decorate walls and educate visitors about better efficiency of such crops in local agriculture. Habitants of the farm and guests are daily served these crops for meals.

The idea of reintroduction of traditional varieties of cereals, pulses and veg-etables is not unique to Navdanya. For example, at the premises of ICRISAT in Hyderabad, I was also served a tasty local millet-based meal for lunch. The only difference is that ICRISAT concentrates on nutritional value and economic inputs of the crop and consider the distribution of old, turned into new crops as a hunger prevention measure, while Navdanya brings the cultural value of these crops, considering them as a cultural heritage.

Dr Shiva also complains of commodification of local cultures and traditions at global level, as she argues, it 'accords with MNCs' interest' (Mies and Shiva, 1993). Local cultures are recognised to have value, but they are fragmented and then 'transformed into saleable goods for a world market' and are used for profit of tourist industry. Such commodification process brings local cultures under one standard and, as a result, it homogenises cultural diversity (Mies and Shiva, 1993, p. 12). A possible way to escape cultural relativism, which is taken by Dr Shiva, unlike Dr Brundtland, as a negative feature, is recognition of diversity and interconnectedness at all levels, between Nature and humans, between different nations and genders (Mies and Shiva, 1993, p. 12).

Using the sustainability rhetoric in public campaigns

On one hand, it is possible to argue that all campaigns of GM debate, both pro- and anti-GMO, use at least a small reference to one or several aspects of the sustainability framework, on the other, there are examples of campaigns which are constructed solely upon the notion of sustainability.

In 2000, the AgBioWorld Foundation was established by two Americans: Professor of genetics Channapatna S. Prakash, based at Tuskegee University, and Gregory Conko, who works as a Senior Fellow and the Director of Food Safety Policy in the Washington, DC-based 'public interest group' called the Competitive Enterprise Institute. The Foundation 'aims to provide science-based information on agricultural biotechnology issues to various stakeholders across the world'. Their 'free electronic newsletter, AgBioView, is a "must read" source of news, research updates and commentary for anyone interested in the latest on advances in plant science, agricultural research and sustainable food production' (AgBioWorld, 2011a).

In 2000, they also ran a campaign to support agricultural biotechnology. They wrote an open letter to the delegates of the eighth session of the United Nations Commission on Sustainable Development, which they accompanied with 'Declaration of Scientists in Support of Agricultural Biotechnology'. The session took place from 30 April through 5 May 2000 and discussed the implementation of sustainable development though its sub-programme and paid a special attention to matters of agriculture (Ecosoc, 2000). Together with other inputs from NGOs and social movements, such as the International Agri-Food Network, the International Federation of Agricultural Producers and La Via Campesina, trade unions and unnamed NGOs, the report also mentioned contribution on biotechnology for sustainable agriculture (Ecosoc, 2000, p. 87).

AgBioWorld has submitted two documents: the first document is an open letter and is signed by six American scientists and three European scientists, among them is Prof. Ammann. In the letter, they write to the UN delegates in anticipation that there would be lobbying from the opponents calling to strengthen the precautionary principle in agricultural regulation and restrict the use of transgenic technologies. Thus, they remind that 'the view that the present day recombinant DNA-engineered organisms pose new or greater dangers to the environment or human health are neither supported by the weight of scientific research nor by a great majority of the scientific community'. They refer to the reports by the US National Academy of Science and a US Congress Committee on Science. Then they move to the sustainability framework, its two aspects environmental and social ones, explaining that GM crops can help reduce use of chemicals and extra land and provide better nutrients for poor farmers. They claimed that transgenic technology could 'help improve people's lives and protect the environment' (AgBioWorld, 2011b, para. 3).

In the accompanying Declaration, signed by 'over 3,400 scientists, including 25 Nobel Laureates such as Dr. Norman Borlaug, Dr. James Watson, Dr. Arthur Kornberg, Dr. Marshall Nirenberg, Dr. Peter Doherty, Dr. Paul Berg, Mr. Oscar Arias Sanchez and Dr. John Boyer' they continued the same rhetoric. GM crops were considered safe for consumers' health and 'environmentally friendly crop plants' that 'preserve yields and allow farmers to reduce their use of synthetic pesticides and herbicides'. Bio-fortified food is mentioned as important products for medical and industrial use. On that basis, they conclude that 'biotechnology can address environmental degradation, hunger and poverty in the developing world by providing improved agricultural productivity and greater nutritional security' (AgBioWorld, 2005).

While the UN report has mentioned the AgBioWorld's contribution, it is not clear how much influence it has actually produced at the UN level. The GMO opponents paid attention to that. For example, LobbyWatch (2016b) mentioned the report in the page about Prof. Ammann. The Foundation itself, however, claimed a widely recognised success by the reference from 'the media as over hundreds of newspapers and magazines, including the *New York Times, Washington Post, International Herald Tribune, Financial Times, Chicago Tribune, Chronicle of Higher Education, The Scientist, Science, Nature*, and *Nature Biotechnology* and *South China Morning Post*', all reporting on this initiative. The Foundation received praise from then Iowa Governor Tom Vilsack of Iowa and the Nobel Peace Prize winner Dr Norman Borlaug, famous for his pro-GM stand (AgBioWorld, 2011a).

Notes

1   This chapter is based on Gerasimova (2016).
2   Losey et al. (1999), argued that Bt corn plants might cause higher mortality rates among larvae of the monarch butterfly. Other scientific research, however, has disproved the previous results and come to conclusion that 'the impact of Bt corn pollen from current commercial hybrid corn on the monarch butterfly population is negligible' (Sears et al., 2001).

Literature

AgBioWorld (2005) Scientists in support of agricultural biotechnology. www.agbioworld.org/declaration/petition.php as viewed 19.06.2016.

AgBioWorld (2011a) About AgBioWorld. www.agbioworld.org/about/index.html as viewed 16.07.2016.

AgBioWorld (2011b) An open letter to the United Nations Commission on sustainable development. www.agbioworld.org/biotech-info/pr/uncom.html as viewed 16.07.2016.

Ammann, K. (2006) Video transcript: Conversations about plant biotechnology. www.condonandroot.com/conversations/v_exp_ammann/exp_ammann.pdf as viewed 26.06.2016.

Ammann, K. (2008) Feature: Integrated farming: Why organic farmers should use transgenic crops. *New Biotech*, 25(2): 101–107.

Ammann, K. (2009) Why farming with high tech methods should integrate elements of organic agriculture. *New Biotech*, 25(6): 3788.

Ammann, K. (2014) Presentation. Asset 2014 GM Debate. Belfast, 9 April 2014.

Antoniou, M. (2014) Question and answer session. Asset 2014 GM Debate. Belfast, 9 April 2014.

Antoniou, M. (2016) Dr Michael Antoniou. www.kcl.ac.uk/lsm/research/divisions/gmm/departments/mmg/researchgroups/AntoniouLab/index.aspx as viewed 16.07.2017.

Antoniou, M., Fagan, J., Robinson, C. (2012) GMO *myths and truths: An evidence-based examination of the claims made for the safety and efficacy of genetically modified crops*. 1st Edition. London: Earth Open Source.

Antoniou, M., Fagan, J., Robinson, C. (2014) GMO *myths and truths*. 2nd Edition. London: Earth Open Source.

Bhatt, V. (2015) Interview with Dr Vinod Bhatt. Dehradun area, 28 February 2015.

Bjerg, O. (2016) *Parallax of growth: The philosophy of ecology and economy.* Cambridge: Polity Press.

Brookes G., Barfoot, P. (2004) *Co-existence of GM and non GM arable crops: the non GM and organic context in the EU.* Dorchester: PG Economics.

Brundtland, G. (2013) ARA Lecture. Technical University. Vienna, 18 November 2013.

Brundtland, G. (2015) Health people, healthy planet: The annual lecture in the business and the environment program. London, 15 March 2015.

Carson, R. (1963) *Silent spring.* London: Hamilton.

Copple, B. (2000) Activist, scientist, yogi? *Forbes,* 30 October 2000. www.forbes.com/forbes/2000/1030/6612054b.html as viewed 16.07.2016.

Davies, S. (2015). Anti-microbial resistance: A global health security threat. Lecture. London: Chatham House. 4 February 2015.

Earth Open Source (2017) Open Earth Source Team. http://earthopensource.org/about-earth-open-source/earth-open-source-team/

Economic and Social Council Official Records (2000) Commission on sustainable development report on the eighth session. Supplement No. 9. New York: UN.

EuropaBio Initiative (2013) *Science not fiction: Time to think about* GM. Brussels: EuropaBio.

Fagan, J. (2007) A science-based, precautionary approach to the labelling of genetically engineered foods. www.psrast.org/jflabel.htp as viewed 23.06.2014.

Fagan, J. (2016) John Fagan, PhD. LinkedIn webpage. www.linkedin.com/in/johnf4 as viewed 16.07.2016.

Gerasimova, K. (2016) Debates on genetically modified crops in the context of sustainable development. *Science, Engineering and Ethics,* 22(2): 525–547.

Giordano, J., Hutchison, P.T., Benedikter, R.A. (2010) Culture, sustainability and medicine in the twenty-first century: Re-grounding the focus of medicine amidst the current 'global systemic shift' and the forces of the market: Elements for a contemporary social philosophy of medicine. *International Journal of Politics, Culture and Society,* 23(1): 29–41.

GMWatch (2010) Open letter to Oxfam-America. www.gmwatch.org/index.php?option=com_content&view=article&id=12130 as viewed 23.06.2014.

GMWatch (2013) Patrick Moore's golden rice campaign featured on the BBC. 6 October 2013. www.gmwatch.org/index.php/news/archive/2013/15101-patrick-moore-s-golden-rice-campaign-featured-on-bbc as viewed 27.07.2014.

GRAIN (2010) An open letter to Oxfam. 12 April 2010. www.grain.org/es/bulletin_board/entries/4221-an-open-letter-to-oxfam-america as viewed 02.04.2014.

Heap, B. (2003) GM crops and the third world, in B.J. Ford (ed.) *GM crops the scientists speak: Proceedings of the 2002 Cambridge conference on genetically modified crops and food, organised by the Cambridge society for the application of research.* Cambridge: Rothay House, pp. 79–87.

ISAAA (2009) *Pocket K No. 3: Are food derived from GM crops safe?* Nairobi: ISAAA.

ISAAA (2016) ISAAA in brief. www.isaaa.org/inbrief/default.asp as viewed 06.07.2016.

ISAAA Brief 44–2012: Executive summary: Global status of commercialized biotech/GM crops: 2012. www.isaaa.org/resources/publications/briefs/44/executivesummary as viewed 26.07.2014.

ISAAA Brief 49: Press release (2014). www.isaaa.org/resources/publications/briefs/49/pressrelease/default.asp as viewed 27.03.2015.

Jackson, W. (1985) *New roots for agriculture.* Lincoln: Nebraska University Press.

James, C. (2014) Global status of commercialized biotech/GM crops: 2014. ISAAA Brief No. 49. Ithaca, NY: ISAAA.

Kos, D. (2012) Sustainable development: Implementing Utopia? *SOCIOLOGIJA*, 54(1): 7–20.

LobbyWatch (2016a) ISAAA: International service for the acquisition of agri-biotech applications. www.lobbywatch.org/profile1.asp?prid=66 as viewed 16.07.2016.

LobbyWatch (2016b) Klaus Ammann. www.lobbywatch.org/profile1.asp?PrId=8 as viewed 16.07.2016.

Losey, J.E., Rayor, L.S., Carter, M.E. (1999) Transgenic pollen harms monarch larvae. *Nature*, 399(214).

Lynas, M., Robinson, C. (2012) Is there a place for GM crops in a sustainable future? *New Internationalist*, November 2012.

Merritt, C. (2003) The status of GM crops across the world, in B.J. Ford (ed.) *GM crops: The scientists speak: Proceedings of the 2002 Cambridge conference on genetically modified crops and food, organised by the Cambridge society for the application of research.* Cambridge: Rothay House, pp. 19–28.

Mies, M., Shiva, V. (1993) *Ecofeminism*. London: Zed Books.

Moore, P. (2014) Golden rice now: Preventing it is a crime against humanity. www.allowgoldenricenow.org as viewed 25.05.2014.

Myers, D., Davidson, J., Chakraborty, M. (1992) *No time to waste: Poverty and the global environment*. Oxford: Oxfam.

Myers, D., Stolton, S. (1999) *Organic cotton: From filed to full product*. London: Intermediate technology.

Navdanya (2016) Navdanya. www.navdanya.org/ as viewed 03.05.2017.

Parrott, N., Marsden, T. (2002) *The real green revolution: Organic and agroecological farming in the South*. London: Green peace Environmental Trust.

Pearce, D., Atkinson, G. (1998) The concept of sustainable development: An evaluation of its usefulness ten years after Brundtland, Working Paper 02, Norwich: CSERGE.

Perry, J.N. (2003) GM crops and the environment, in B.J. Ford (ed.) *GM crops: The scientists speak: Proceedings of the 2002 Cambridge conference on genetically modified crops and food, organised by the Cambridge society for the application of research.* Cambridge: Rothay House, pp. 43–66.

Richardson, D. (2001). A farmer's view. In *GM crops, understanding the issues*. London: UK Biotechnology Industry.

Salazar, C. (2010) Vandana Shiva: The elusive search for sustainability. *This Is Africa, News*, 17 September 2010. www.thisisafricaonline.com/News/Vandana-Shiva-The-elusive-search-for-sustainability as viewed 16.07.2016.

Sears, M.K., Hellmich, R.L., Stanley-Horn, D.E., Oberhauser, K.S., Pleasants, J.M., Mattila, H.R., Siegfried, B.D., Dively, G.P. (2001) Impact of Bt corn pollen on monarch butterfly populations: A risk assessment. *Proceedings of the National Academy of Sciences*, 98: 11937–11942.

Shiva, V. (2001) Special report: Golden rice and neem: Biopatents and the appropriation of women's environmental knowledge. *Women's Studies Quarterly*, 29(1/2): 12–23.

Shiva, V. (2011) Introduction, in V. Shiva, D. Barker, C. Lockhart (eds.) *The GMO emperor has no clothes: A global citizens report on the state of GM. false promised, failed technologies*. Florence: SICREA.

Shiva, V. (2012a) Foreword, in A. Whitney Sanford (ed.) *Growing stories from India: Religion and fate of agriculture*. Lexington: Kentucky University Press.

Shiva, V. (2012b) On the problems with genetically modified seeds. Interview with Bill Moyers. 13 July 2012. http://billmoyers.com/wp-content/themes/billmoyers/transcript-print.php as viewed 23.06.2014.

Shiva V. (2013) Seed Monopolies, GMOs and Farmer Suicides in India – A response to Nature. Response by Dr Vandana Shiva to an article published on 1st May 2013 in *Nature* by Natasha Gilbert. http://www.navdanya.org/blog/?p=744 posted 12 November 2013.

Shiva, V. (2014) *The Vandana Shiva reader*. Lexington: Kentucky University Press.

Swaminathan, M. (2003) Strategies towards food security. *Social Scientist*, 31(9–10): 58–94.

Tripp, R. (2006) *Self-sufficient agriculture: Labour and knowledge in small-scale farming*. London: Earthscan.

Tripp, R. (ed.) (2009) *Biotechnology and agricultural development: Transgenic cotton, rural institutions and poor-resource farmers*. London: Routledge.

UNEP (2010) *What is biodiversity?* Geneve: UNEP.

United Cities and Local Governments (2011) *Culture: Fourth pillar of sustainable development*. Barcelona: Institut de Cultura. www.agenda21culture.net/index.php/docman/-1/393-zzculture4pillarsden/file as viewed 16.06.2017.

Visvanathan, S. (1991) Mrs Brundtland's disenchanted cosmos. *Alternatives*, 16: 378–381.

WB (1994) Report N 13698. Project Completion Report. India. National Social Forestry Project. Credit 1611-IN. 10 November 1994. New York: WB. www-wds.worldbank.org/external/default/WDSContentServer/WDSP/IB/1994/11/10/000009265_39610071526 18/Rendered/PDF/multi_page.pdf as viewed 16.07.2016.

WCED (1991) *Our common future*. Oxford: Oxford University Press.

WFP (2002) *United nations statement regarding the use of GM foods as food aid in Southern Africa*. Rome: WFP. http://documents.wfp.org/stellent/groups/public/documents/newsroom/wfp076534.pdf?_ga=1.43592805.1115906955.1468660135 as viewed 16.07.2016.

WHO (2002) WHO Director General speaks on GM foods. www.afro.who.int/en/mediacentre/pressreleases/item/457-who-director-general-speaks-on-gm-foods.html as viewed 16.07.2016.

6 The alterglobalist discourse

Introduction: what is globalisation and alterglobalist movement?

Another important discourse in the debate on GM crops is alterglobalism. To analyse the alterglobalist discourse, we start with an understanding of what globalisation is. Globalisation is the 'big idea' of the late twentieth century with many definitions, and it faces the danger of becoming 'the cliché of our times' (Held et al., 1999). First, the term 'globalisation' was introduced in French language in 1904 and then appeared in English in 1930. Spooner, however, argued that the globalisation process started to accelerate only in the 1990s, after a static bipolar system with the fall of the Soviet Union forced a re-evaluation and the consequences of globalisation processes became more evident. It points to a complex set of global processes. On one hand, there was a 'progressive expansion' of diverse interactions at global level, including the spread of innovations and new technologies. On the other, 'rules have begun to fade and our tolerance of uncertainty has risen' (Spooner, 2015, pp. 1–3).

In a very general definition, globalisation is an 'accelerating change' (Spooner, 2015, pp. 1–3). It is also 'the widening, intensifying, speeding up, and growing impact of world-wide interconnectedness . . . across all key domains of human activity, from the military to the cultural' which is characterised by four types of change: extensity (i.e. stretching geographical scope and transcending national boundaries, intensity (i.e. deepening the level and degree of interconnectedness), velocity (i.e. speeding global processes and interactions in every sphere) and impact or which can also be referred here as glocalism (i.e. local and global characteristics and events produce a significant influence on each other) (Held et al., 1999).

Three broad accounts of the nature and meaning of globalisation can be identified, referred to here as the hyperglobalist, the sceptical and the transformationalist views. These define the conceptual space of the current intensive debate about globalisation (Held et al., 1999). As can be seen, many activists within the alterglobalist movement arguing for radical change in global polity belong to the third category.

Globalisation is often analysed in the context of the modern system of neo-liberal capitalism, based on the free-market system and global expansion of international

trade and foreign investment, as defined by Marxism. Thus, globalisation in relation to the global distribution of capital, in developing countries particularly,
represents new relations of agricultural production, social reproduction and global
power relations 'infused with resistances' (McMichael, 2007). The anti-globalist
resistance poses the question of the rethinking of the role and identity of agrar-
ians, referred as to 'new peasantry' and 'agrarian citizenship' (McMichael, 2007).
Colonialism and post-colonialism is also a part of this process, as globalisation is
accompanied by Western expansionism and cultural imperialism.

Globalisation is first of all understood as a dominant political discourse which
questions the power of a national state (Trouillot, 2003). A profound change that
globalisation has brought is the challenge to social actors to adapt to the new global
institutional environment. Thus, it can be argued that social activism against globalisation is a part of the global search among different actors to adopt a working
strategy in the globalised world to better represent their group's interests and rede-
fine their identity and values.

Kahn and Kellner argued that the 'worldwide resistance to globalization policies' is one of the most important political developments in modern history. Such
resistance must 'be understood as pertaining to highly complex, contradictory, and
sometimes ambiguous varieties of struggles that range from the radically progressive to reactionary and conservative' (Kahn and Kellner, 2007, p. 662).

Thus, anti-globalisation is both a critical rhetoric of globalisation and a network of social movements. By questioning whether the anti-globalisation movement is a movement, Klein pointed out that it is only against a specific set of
features of globalisation and that 'it is more accurate to picture a movement of
many movements – coalitions of coalitions' (Klein, 2004, p. 219). The movement
is very heterogeneous, it is reflected in its synonyms of alter-globalisation and
counter-globalisation. Its different anti-globalisation critiques and campaigns are
explained by different definitions and understandings of globalisation: for example, some see it as a concrete economic and political project by governments,
business, elites and others – as a characteristic of the neo-liberal capitalist system
(Juris, 2008, p. 7).

This then leads to different social groups aiming at different end goals and
tactics. Crawford (2005) suggested a loose distinction of six categories within the
anti-globalisation movement: environmental and social justice movements, third
world groups, organised labour groups, the indigenous rights movement, national-
ist groups and the moral majority movement, acknowledging that often groups
can have ideologies common to different categories. Competition or struggle for
influence among groups produce different 'ideology (antiglobalization versus anti-
capitalism), strategies (summit hopping versus sustained organizing), tactics (violence versus nonviolence), organizational form (structure versus nonstructure),
and decision making (consensus versus voting)' (Juris, 2008, p. 15).

The anti-globalisation movement first appeared in global media with organised
mass protests against institutions such as the WTO, International Monetary Fund
(IMF), World Bank, World Economic Forum, and the Group of Eight (G8). These
have offered their own response and alternatives, such as the World Social Forum,

to counter-events responding to such leading summits and annual meetings of the mainstream global political system.

One of the most numerous protests which also attracted media attention was that against the WTO Ministerial Conference of 1999 in Seattle, US, also known as The Battle of Seattle. It united more than 40,000 protestors and some activists, such as Jose Bove, saw it as a revival of protest movements in the tradition of that of 1968 (Bleiker, 2002). Thus, references to the Seattle marches are very common in this discourse.

These new protests reinvigorated the debate over globalisation and have also included discussions of the role of GMOs, corporations and global politics. While the GM crops debate can directly or indirectly relate to any of the six previously cited categories. From the analysis provided in this chapter, it can be suggested that in the discourse of alterglobalist activism there are three sub-discourses that criticise GM crops: the sub-discourse of systematic change which often takes an opposition against global corporate business, the new leftism brought by the global peasant movement and the regionalism versus globalism sub-discourse based on cultural and culinary capitals.

The alterglobalist search for alternative development and systemic change

This sub-discourse is based on the fundamental assumption that the social order taking place today is built upon dominance of global capital, global industrial complexes and ideas 'inspired by neoliberal visions of nations as resource pools and open markets operating without restrictions' (Best, 2011, p. xviii). These complexes include military, media, medical, and so forth and agricultural ones. Thus, neo-liberal capitalism has become 'the new paradigm of permanent growth' without acknowledging the needs of the majority of global citizens (Best, 2011, p. xxi). And there was need to struggle against it:

> Surrender . . . is not an option . . . the only solution lies in organizing informed radical change across all levels of the integrated systems of domination – commencing with an emancipatory education into and critical understanding of the precise nature and dynamics of the systemic barriers blocking our journey into sustainable planetary community.
>
> (Best, 2011, p. xxiv)

Naturally, anti-GM activists, members of the alterglobalist movement, pay most attention to the agricultural complex. Dr Vandana Shiva and Dr Walden Bello are the most well-known activists in the alterglobalist discourse who search for alternative model of development. Shiva and her views have already been introduced in Chapter 3, while Dr Walden Bello is 'the most respected anti-globalisation thinker in Asia', according to *Le Soir*. Born in Manila, the Philippines, in 1945 Dr Bello is 'a human rights and peace campaigner, academic, environmentalist and journalist'. He holds a PhD in Sociology from Princeton University. In parallel to Vandana Shiva, he has been active in setting up and running NGOs. In the 1980s, he joined the NGO Food First in the US and served as chairman of the board of Greenpeace Southeast Asia. For his activism, he was awarded the Right Livelihood 'Alternative Nobel Prize' with three others for 2003 (Asia Institute, 2003). He co-founded an NGO called 'Focus on the Global South' in 1995 with UNDP Senior Advisor Kamal Malhotra. The organisation has offices in Bangkok, New Delhi and Manila and lists 30 international and regional civil society organisations, including Friends of the Earth, Third World Movement and La Via Campesina (Focus, 2016a).

'Focus' was set up in 1995 'to challenge neoliberalism, militarism and corporate-driven globalisation while strengthening just and equitable alternatives' (Focus, 2016a). These alternatives include building and supporting international networks and process – for example, Our World is Not for Sale (OWINFS) and the World Social Forum promotion of the concept and term 'deglobalization' by Bello; '"derailing" the Doha round of WTO talks'; allying with other social movements that resist globalisation regionally and internationally; producing books, newsletters and other media; 'regaining alienated commons', especially in terms of land reform and rights to public water (Focus, 2016b).

The logic employed by both activists' works on globalisation is the same: that global capitalism does not work, as indicated by multiple crises. Meantime, the global resistance began with the Seattle protests aiming to provide an alternative to such capitalist regimes.

Shiva has identified three major global crises: climate, energy and food, which result from industrialisation of agriculture and the effects of globalisation and trade liberalisation of agriculture (Shiva, 2011, p. 169). She put globalised agriculture in the centre of global crises. Bello agreed, saying that 'capitalism seems poised to fully subjugate agriculture, its dysfunctional character is being fully revealed' by a series of global crises in economy and food production (Bello, 2009, p. 36).

For his part, Bello has analysed global crises, particularly in food commodity price fluctuations and the rise in chronic hunger as a result of 'the massive agricultural policy reorientation known as 'structural adjustment'" (Bello, 2009, p. 6). He criticised the mainstream economic analysis of dynamics which drive the global food demand as offered by some economists including the distinguished British economist, Paul Collier. In fact, both Bello and Collier agreed that the food crisis of 2008 was one of production. But while Collier argued that increased global food demand is driven by prosperity and that production could not satisfy that demand, Bello saw the root of the crisis in replacing small-scale agriculture with global agrofirms producing for global demand (Bello, 2009, pp. 7, 37). The most famous agrofirm that he talks about is, of course, Monsanto, which delivers 'false promises' and represents the market-centred 'against people and planet' paradigm (Shiva, 2011, pp. 180–188).

In 2009, Bello assumed that global crises and 'the collapse of the global real economy' would lead to a system reverse in the form of 'deglobalization' and a refusal of global citizens be led by capitalism (Bello, 2009, pp. 36–37). He assessed global capitalism as hardly accountable and responsible for environmental and social crises. Bello, in the Marxist tradition, distinguished two stages in development of 'the international agrifood system'. The first started with the emergence of the first regime at the end of the nineteenth century based on the colonialism of the Global South and reorientation of its food production towards a global market. The second is the post-war Bretton Woods regime which kept protectionist regimes for key world powers in agriculture, such as the US, but allowed for the expansion and establishment of a 'hegemony of corporate industrial agriculture through the institutionalisation of free trade rules and monopolistic property rights favouring the spread of globally integrated production chains' (Bello, 2009, p. 38). The next point after identifying the global crisis is to identify an alternative model of development.

Shiva wrote that 'we can and must respond creatively to the triple crisis' (Shiva, 2011, p. 169). From that point, the Seattle event was most welcomed and seen as 'a historic watershed', at which point Shiva hoped it was proof that globalisation is not 'an inevitable phenomenon which must be accepted at all costs but a political project which can be responded to politically' (Shiva, 1999). As the Seattle manifestations protested against a new round of trade negotiations which did not included the interests of developing countries, Shiva also saw it as part of post-colonial discourse (Shiva, 1999).

Seattle itself was symbolic since it hosts such corporations as Boeing and Microsoft. However, the first round of anti-globalisation protests was against international organisations, not corporations which remained in their shadow: with corporations remaining in the background, the proponents of free trade were going out of their way to say that the WTO was a 'member-driven' institution controlled by governments that made democratic decisions. But the Seattle conference which excluded developing countries showed that global governance was a non-transparent and anti-democratic processes (Shiva, 1999). Shiva also offered a new name for WTO – World Tyranny Organization. It is referred to as a tyranny because it supported a global regime with destructive social and ecological impacts and endorsed the further impediment of corporate control on food, the environment, work and the future of humanity (Shiva, 1999). Bello also saw the WTO 'as the lynchpin of a multilateral system of economic governance that would provide the necessary rules to facilitate the growth of global trade and the spread of its beneficial effects'; as a 'robbery' because of its agreements, including the Trade-Related Investment Measures (TRIMS), and the Trade-Related Aspects of Intellectual Property Rights (TRIPS) not acting in the interests of developing nations (Bello, 2000).

He criticised international agreements on agriculture which led to developing countries signing 'away their right to use trade policy as a means of industrialization'. It allowed MNCs such as Novartis and Monsanto '[to] monopolise hi-tech innovations and monopolised interaction with nature', so that developing nations had to open up their markets into which corporations dumped their production surpluses and destroyed smallhold agriculture. The principle which illustrates these processes, which Bello cited from Ralph Nader, was 'trade above all' – trade 'uber alles' (Bello, 2000). WTO was portrayed as a global trade system led by global corporations. Under these conditions, corporations began to move out of the shade of the 'private penumbra', and he argued that it was possible 'to effectively crystallize opposition', finding a great opportunity for the global resistance movement.

The Seattle protesters were students, farmers, workers and environmentalists, indicating the heterogeneous character of the alterglobalist movement. Seattle was as important to Bello as it was to Shiva, since it brought together diverse people who 'were united by one thing: their opposition to the expansion of a system that promoted corporate-led globalization at the expense of social goals like justice, community, national sovereignty, cultural diversity, and ecological sustainability' (Bello, 2000). The clash of protesters with police during this event which had started peacefully indicated 'the intolerance of democratic dissent, which is a hallmark of dictatorship, was unleashed in full force' (Shiva, 1999).

The protests which followed after Seattle were demonstrations against the World Economic Forum. Walden Bello saw it as 'a critical mass to turn the tide against corporate-driven globalisation'. His main point of critique was a reliance of global system on neo-liberal principles such as free trade and transnational corporations:

> The unrestricted flow of goods and capital in a world without borders was said to be the best of all possible worlds, though when some observers pointed out that to be consistent with the precepts of their 18th century prophet, Adam Smith, proponents of the liberal doctrine would also have to allow the unrestricted flow of labour to create this best of all possible worlds, they were, quite simply, ignored.
>
> (Bello, 2000)

He used the rising poverty in the world as an illustration of the neo-liberal regime's imperfections. The production paradigm argued by economists such as Paul Collier and Henry Bernstein, who supported the globalisation of food production, is global industrialised farming and food production or 'the global market-driven paradigm'. In turn, Bello offered 'a local-market-centred paradigm' as an alternative (Bello, 2009, p. 5).

One sector was viewed as particularly dangerous to the local-market-centred model: genetic engineering, which was seen as 'wrestling almost complete control of the actual physical process of production from [smallholder farming] and promised to correct the diseconomies of large-scale production that had allowed the family farm to survive' (Bello, 2009, p. 33). Biotechnology, he argued, has contributed to the spread of global capital by including new organisms in the input commodity production of food and a vertical integration of a global firm (Bello, 2009, p. 33).

Bello argued that the rise of biotechnology within the food industry has only strengthened such a global market-based production model, removing the farmer from production, 'accelerate[d] his or her dispossession and conversion into a rural worker' (Bello, 2009, p. 38). Bello dismissed the argument that smallholdings are less productive and innovative than large-scale farming, as posited by Collier, for example. He instead cited van der Ploeg, who wrote that 'technology is not only about linking artefacts and governing material flows – it is much about interlinking people in specific ways in order to obtain the rights kind of conditions and flows' (van der Ploeg, 2008, cited in Bello, 2009). In that sense, big business can produce only a 'standardized outflow', while small-scale farming can better deal with 'specificity or variation' and 'advanced science and peasant agriculture' should not be in contradiction but rather be incorporated into one another (Bello, 2004, p. 142). On the other side, GM seeds were enriching corporations and extending production chains. With 'terminator seed' biotechnology, he argued, 'could in fact provide the death blow to the peasantry' (Bello, 2004, p. 33). Bello reflected upon technocracies and referred to Kuhn's idea of cycles of scientific development, pointing to a modern scientific paradigm crisis, the main root of this crisis lies in the process of corporate-led globalisation (Bello, 2000).

Bello identified GM seeds as a tool for competing over global influence, not just in developing countries by Western nations but also among themselves. He recognised that Western countries compete for global influence and referred to 'a strong corporate lobby' that believes that the environmental concerns of GM crops are imposed by Europe as part of a competitive strategy (Bello, 2004, p. xviii).

At the time of Seattle, the search for an alternative to global rough capitalism was popular. Even British Prime Minister Tony Blair articulated the vision of 'compassionate globalisation', a 'Third way' providing 'a new alternative . . . on the centre and centre-left' of politics but with 'clear values of social justice, democracy, cooperation' which was a combination of the liberal commitment to free trade and the social democratic commitment. It was denied by alterglobalists such as Shiva and Bello (Bello, 2000).

Shiva, as shown in Chapter 3, has argued for local economy and small-scale agriculture, free from GM technology. Bello too hoped that the search for alternatives would bring back 'a sympathetic look' to small-scale agriculture (Bello, 2009, p. 37). He disagreed with economists such as Collier that the 'New Green Revolution' was the solution. He dismissed the argument that GM crops could bring benefits to developing countries and be part of a solution to global food crises to provide for food shortages (Bello, 2009, pp. 9–10). His alternative paradigm is based upon concepts of food sovereignty and the values of global peasantry instead.

Global peasantry, regionalism versus globalisation and new French leftism

Another iconic figure within the alterglobalisation movement is Jose Bove. His eccentric behaviour attracted much media attention and made him the perfect ambassador to communicate the ideas of the movement. His ideas and campaigns represent syndicalism, resistance and locality. His looks consisting of moustache, pipe and an Asterix-like appearance made him stand out from the crowd and be favoured by the media (Morena, 2013). He is often idealised by the media since he is seen as a man of the earth armed with only courage to face the power of industry, economics and the political, a modern day David versus Goliath hero whose phrase, 'No passarant' ('they won't pass'), flatters the French national revolutionary vanity and explains his popularity (Kaci and Kerhuon, 2007, p. 18).

Hailing from southern France, agriculture was part of his family tradition, since his great uncle, Nicolas Bove, was a French botanist who managed the gardens of Ibrahim Pasha in Cairo (Kaci and Kerhuon, 2007, p. 21). Joseph (Jose) Bove was born in Bordeaux in 1953. His parents also worked in agriculture, being employed by INRA (National Institute for Agricultural Research) and had spent some years abroad in the US (Aries and Terras, 2000, p. 6). At one point, the media made allegations that Bove's father, Josy, was 'an ardent support of genetically modified organisms', criticising his son for demonising GMOs in the same style as people involved in magic were burned in the Middle Ages in an effort to present an ideological conflict between father and son. The case was put to rest by a public article issued by Josy Bove in *Le Monde* wherein he made clear that he had never worked on GM crops (Pingaud, 2002, pp. 25–27).

Bove developed an interest in international worker and peasant movements while reading world famous authors on political economy such as Henry David Thoreau, Gandhi, Bakunin and Proudhon (Pingaud, 2002, pp. 25–27). In 1971, he engaged in the anti-militarist movement and took part in a protest against the extension of the military camp in the Larzac which also concerned farmers residing in the area. After the Larzac demonstrations, Bove chose to become a farmer himself and settled on a farm in the Pyrenees, near Béarn (Bove and Dufour, 2001, pp. 36–37). The Larzac demonstrations also mobilised activists and farmers, leading to the creation of a foundation named Larzac Solidarites which provided a platform for maintaining a network of people and offered financial support for further campaigns including those against McDonalds and the GMOs (Terral, 2011).

Thus, Bove has arguably participated in all stages of the development of global protest and new peasant movement: the first stage being the crystallisation of the movement (1971–1981); from 1981–1999, the reinvigoration of the peasant movement; and the period beginning after the year 2000 covering anti-globalisation rhetoric (Terral, 2011, pp. 21–22). In 1987, after having already had his 'syndicalist' experience in the Confédération Nationale des Syndicats de Travailleurs Paysans (CNSTP) and criticising the Fédération Nationale des Syndicats d'Exploitants Agricole (FNSEA), Bove co-founded a new organisation named the Confédération Paysanne (CP) in order to adopt 'a new approach to agriculture to replace intensive farming' (Bove and Dufour, 2001, p. 48). It came as a result of farmers' meetings jointly organised by the FNSP (National Federation of Farmer Unions) and the CNSTP. They aimed to offer a space for the expression of ideas to peasants opposed to the policy of aiming at agricultural productivity and against smallholders as advocated by the FNSEA and the Centre National des Jeunes Agriculteurs (CNJA) (CP, 2016).

According to its recent website, the aim of the farmers' confederation is 'to fight for an agriculture that respects the income of peasants, independent agro-processing, distribution, integrations . . . open to alternative practices, meeting the expectations of the society, innovative and attentive to the rural life' (CP, 2016). It supports sustainable agriculture and campaigns against 'the programmed disappearance of peasants'. Its founders believe that their work is not in vain as they report on 'twenty-five years of struggle on all fronts', including against industrialisation of agriculture and in favour of biodiversity conservation, social solidarity and social mobilisation (CP, 2016).

His famous campaign on food was begun by Bove in 1999 when the US and the European Union were experiencing tension over food exports. The EU banned the import of American hormone-treated beef, and the US responded by adding customs surcharges on European food products, including cheese. In the South Aveyron region of France, most farmers produce ewe's milk, which is used in the production of cheese, particularly Roquefort. As a result of the surcharge, the price doubled, effectively leading to a slump in sales and farmers losing '15 million francs' (Bove and Dufour, 2001, pp. 4–5). Bove, as a farmer himself, was affected by this US policy.

Farmers of the region got together, printed pamphlets and banners saying 'No to the US embargo on Roquefort' and distributed around villages, reached out for media coverage and met with the minister of agriculture who informed farmers that there was 'nothing he could do, that he was unable to obtain direct financial compensation, that Europe was powerless and there was no other way out'. As they found lack of support, they had no option but 'to take on McDonalds, the symbol of industrial food and agriculture' (Bove and Dufour, 2001, pp. 4–5). The protest against the restaurant chain was conceived as 'a non-violent but symbolically forceful action' (Bove and Dufour, 2001, p. 5). The demonstrators dismantled pieces of the roof, doors and windows of the unfinished McDonalds building in Millau, France. Five farmers' leaders were arrested as a result of the rally. Bove was also imprisoned. The arrest of the farmers mobilised the public with a petition to support both the ewe milk producers and the incarcerated farmers (Bove and Dufour, 2001, p. 10). At this moment, Bove realised the importance of the media and its influence on public opinion:

> The handcuffs symbolized my arrest. I realized the impact that the image of my holding them up could have, so the photograph was not accidental. . . . It certainly helped to extend the mobilization and underline the fact that a legitimate protest movement could not be stifled.
>
> (Bove and Dufour, 2001, p. 11)

He had realised the power of public fear. While the issue of hormone-fed beef had occurred appeared in the 1980s, it was 'overnight' that Bove and his fellow farmers 'realized that globalization was forcing us to eat food that contained hormones'. The same fear has brought about the initials 'GMO' (Bove and Dufour, 2001, p. 12).

The campaign about French cheese mobilised people from trade and farmers' unions because 'Roquefort [was] crucial to the local economy' (Bove and Dufour, 2001, p. 4) since the milk processing industry provided jobs.

French farmers also received support from abroad, such as letters from American farmers and even more than 30,000 francs after the Millau campaign donated. The new French movement also made contact with La Via Campesina and the European anti-GMO movement (Bove and Dufour, 2001, p. 31). In 1998, CP co-founded ATTAC (Action for a Tobin Tax to Assist the Citizen), a powerful social justice NGO (Morena, 2013, p. 104).

The anti-GMO rhetoric had begun to attract Bove's attention in the early 1990s. At that time, he understood that it would be very difficult 'to muster up an effective opposition to the pharmaceutical giants' and it would be more productive to act in coalition with other activist groups. As a result, Bove co-founded the Farmers', Ecologists' and Consumers' Alliance (Bove and Dufour, 2001, p. 82). Bove went on to personally participate in protests against GM crops: in 1998, he destroyed a stock of GM maize in Nerac, while in 1999, he destroyed a GM trial rice crop in Montpellier and visited Seattle to join in demonstrations against 'Frankenstein food' (Bove and Dufour, 2001, p. 84).

Bove is against GM seeds being used in agriculture; as a farmer, he found them unnecessary – 'We don't need GMOs to do our job [in agriculture]'. He also believes that transgenic agriculture enriches MNCs that enforce industrial patents and are paid 'royalties for life itself' (Bove and Dufour, 2001, p. 85). His opposition to GM crops is based upon environmental, scientific and political concerns. The highest risk that he sees from GM crops is the 'irreversible effect on biodiversity'. He refers to the example of the Monarch butterfly, a species allegedly suffering as a result of GM activity (Bove and Dufour, 2001, p. 86). He calls genetic modification 'a technique of tyranny', part of the American hegemony, and explains the spread of GM maize, which also happened to be the first industrial crop of the US (Bove and Dufour, 2001, p. 89). In addition, he supports the terminator gene argument:

> The technique introduces a gene that stops the grain from germinating once it reaches maturity. It makes perfect commercial sense, ensuring a hundred-per-cent return on investment. The seed growers economize on the lawyers' fees that they would otherwise have incurred prosecuting farmers for theft.
> (Bove and Dufour, 2001, p. 90)

Bove has stated that he does not believe that GM crops can be used to alleviate hunger (Bove and Dufour,2001, p. 95). The position adopted by French politicians considering authorisation of commercial GM crops was criticised by Bove, who referred to it as a 'very short-sighted, mercenary, and conformist viewpoint – apparently based on scientific evidence, but actually without any such thing' (Bove and Dufour, 2001, p. 92). He also spoke against the agricultural research centre, CIRAD, located in Montpellier and funded by the French government on the basis that 'they [CIRAD] couldn't claim to speak for independent scientific research while at the same time they worked for agribusiness'. To challenge this scientific research, Bove allied with a group called 'GMO Watchdog' to produce a report entitled 'Researchers, come out of your labs' to educate farmers about GMOs (Bove and Dufour, 2001, p. 93). He is also convinced that 'scientists are playing around with genetic modification without knowing the real nature of the genome' and act very much like 'apprentices' (Bove and Dufour, 2001, p. 98).

To overcome all these challenges, Bove offered 'multifunctionism', which is a social model balancing social forces, a social order under which consumers respect biodiversity and farmers. This order avoids 'accepting intensive agriculture' run by a small group of farmers with largeholdings (Bove and Dufour, 2001, p. 125). Sustainable multifunctional agriculture is not 'reduced to mere trade'; it combines both concepts of food security in its regional production sense and food safety as defined by consumers (Bove and Dufour, 2001, p. 163).

In 2001, Bove remained optimistic about the outcome of his struggle and credited the EU ban on GM crops as achievement of his syndicalist activities:

> Our experience of struggle all over France – from our campaign for fairer land distribution through our opposition in the 1970s to the extension of the military camp in Larzac, to the recent fight against GMOs and junk food – has shown us that we can be successful. On the issue of GMOs there have been huge advances, the latest in January 2000 when a protocol, agreed in Montreal by one hundred and thirty countries, adopted a precautionary principle on the import of GMO products.
>
> (Bove and Dufour, 2001, p. 169)

In 2007, Bove ran in the French presidential election. He presented his candidature as 'alternative left', representing millions of citizens who suffer from social insecurity, who no longer believe in traditional left or right wing and pledging to fight against technocracies and the neo-liberal economic regime (Bove, 2007, pp. 7, 9). It is hardly surprising that in his political agenda, which consisted of eight principles, he included the question of GM crops and nuclear research, which he suggested be decided by a citizens' debate conducted democratically and fully transparent (Bove, 2007, p. 177).

Although he did not win, he moved forward with his political career, being elected as a parliamentarian representing the Greens/European Free Alliance in the European Parliament. There, he continues to defend his values of farming and anti-GM rhetoric. For example, in August 2015, he contributed to the debate on GM crops in Europe that was held in the European Parliament. He raised three points: first, that GM crops still needed to be risk assessed in the areas of science, health, environment and socio-economic development. Second, he pointed out that the majority of the Europeans are against GMOs and then reminded parliament about the obligation of authorities towards workers within the dairy and animal industry regarding the authorisation of the use of GM feed (Bove, 2015).

Bove and his syndicalism is well covered in the literature, since his ideas and campaigns represent a good case for discussing how the new French left politics and alterglobalism are connected. As seen in the course of Jose Bove's political career, the social movements of the 1980s and 1990s were central to the construction of the 'alter globalization' global justice movement which came as 'the current wave of global protest' (Fletcher et al., 2013, p. 1). Bove rose on a wave which brought new values of the solidarity and egalitarism that are traditionally close to the French heart, and it is for this reason that he gained popularity. In the late twentieth century in France, as a result of Bove's campaigns, farmworkers began to symbolise the values of autonomy and authenticity in the context of growing distrust towards national and international institutions.

There was also a shift in agricultural syndicalist power: agricultural syndicates such as the FNSP started to lose their popularity to new cooperatives such as the CP which represent 'the new left wing peasant' (Morena, 2013, p. 101). The CP's strategy of making alliances and co-founding organisations with different movements working on agricultural, social justice and environmental issues, particularly La Via Campesina, allowed the organisation to reach out to different groups and make themselves widely known. The use of the legacy of anarcho-syndicalism, particularly a decentralised form of organisation and open membership have also enabled it to, on the one hand, 'dialectically [retain] anarchist values' while still attracting a vast majority of members who although not anarchists were attracted to autonomy within the organisation's management and its open membership (Heller, 2012, p. 71).

With opposition to traditional syndicates such as FNSP and right wing politicians such as Nicolas Sarkozy, who was Bove's opponent during the presidential elections, the alterglobalist movement in France has become the political alternative to the right wing and has raised new issues. They are economic matters and the issues of national identities which were embodied in the values of authenticity and tradition of French agriculture and cuisine (Waters, 2012). Such values can summarised under the concept of culinary capital, and it appears that France is not the only European country to use culinary capital to protect the interests of its agricultural community. Italy has made a similar argument and culinary capital, regionalism versus globalisation is used to promote local agriculture and food industry. GM crops, since they are connected to global capital and American MNCs, go against the values of regionalism and are thus actively opposed and rejected. It is also argued that CP and its leader, Bove, were able to succeed with the GM debate in France because they used the rhetoric of solidarity and syndicalism which focused on the small farmer who has traditionally enjoyed support from the nation (Heller, 2012).

Regional cuisines, culinary capital in the debates on globalisation and GM crops

The concept of culinary capital references the Marxist idea of capital and Bordieu's explanation of how multiple forms of capital allow people to acquire status and power (Calhoun and Sennett, 2007, p. 7). Food in social terms goes beyond its nutritional value and becomes a system for communication, as it serves to express personal and group identities and even 'cement social bonds' (Belasco, 2008, p. 15). Thus, culinary capital is a reflection of the economic and class hierarchy within society.

Certain foods and eating practices confer status and power but also reflect cultural heritage and the preservation of regional agricultural and food practices. The modern appearance of gourmet food culture or the 'gourmet-ification' of culture has brought with it two notions – authenticity and exoticism within the food discourse (Johnston and Baumann, 2010).

Culture also shapes consumer choices, and eating has become connected to civic activism. Naccarato and Lebesco suggested using 'good eating practices . . . [as] an indicator of being a good citizen' in self-identification. It also allows space for empowerment and even political resistance (2012, p. 5). Another recent term is 'foodie culture', which describes a particular interest in tasting different cuisines and the growing appreciation of food and culinary capital. There are many different sub-discourses within foodie culture: adventurism and omnivorousness, 'Irrational food phobias', elitism and a form of populism produces antagonism between the promotions of 'privileged culinary practices that prioritize sustainability, health, and dietary restraint' (Naccarato and LeBesco, 2012, p. 11).

Thus, cuisine becomes a language that helps society and citizens to establish their identity, and 'counter-cuisine' is a new tool enabling the challenge of mainstream values and political trends via food (Belasco, 1989). For example, Julia Child used her cooking recipes of French cuisine to establish her upper class credentials but also hide the economic realities of class inequalities through the assumption that French cuisine was accessible to everyone (Naccarato and LeBesco, 2012). The authors have also offered another useful term which describes civic activism in food – 'culinary dissent' (Naccarato and LeBesco, 2012, p. 19). That is precisely what Bove has been doing in his campaigns. At his protest against McDonalds, he came up with the term 'malbouffe' which translates as junk food and refers to not only the aspects of food safety but also issues of taste and authenticity. This is how Bove explains it:

> The term means both the standardization of food like McDonald's – the same taste from one end of the world to the other – and the choice of food associated with the use of hormones and GMOs, as well as the residues of pesticides and other things that can endanger health. So there's a cultural and a health aspect. Junk food also involves industrial agriculture – that is to say, mass-produced food.
>
> (Bove and Dufour, 2001, p. 54)

Another example of how culinary capital is used to promote and protect local agriculture and food industry can be found in Italy. Italian cuisine is also an interesting example of how the values of regional cuisines has passed borders but maintained its values.

In 2013, while on a flight to Verona, I spoke to an Italian lady sitting next to me. As soon as she learnt that I was researching political perceptions of GM crops, she told me that Italy did not need them, that the beauty of Italian cuisine has come from the freshness and organic nature of the vegetable being used. A similar approach can be found among many Italian chefs promoting Italian cuisine, such as Gino Barbaro, an Italian chef living in the US (2013) who is 'a big proponent of buying and growing organic food', and who has written book on how to grow vegetable organic manner and composting (Barbaro, 2013, p. 28). Similarly, Antonio Carluccio, another Italian chef living abroad – in the UK – has argued that Italian cuisine makes use of the best of ingredients: 'minimum of fuss, maximum of flavour' with 'no need for artifice or glamorising if the basics are right' (Carluccio, 2012, pp. 2–3).

Under this logic, local and organic agricultural crops are contradictory to GM crops. 'Every instinct I possess about food and nutrition goes against genetic modification of any kind', said Carluccio in an interview. He reminds the reader about the importance of taste, which comes from ingredients that are locally produced in the traditional manner: 'People are more aware now about how science can step in and produce food intensively, with a good-looking appearance and longer shelf-life, but not necessarily the best flavour' (Carluccio, 2001).

Carluccio went further – from writing, he moved into activism. He joined together with three other food writers in the anti-GM campaign organised by Greenpeace in 1999 to secure a ban on the release of all GM organisms into the food chain (Watson-Smyth, 1999). The chef also developed, together with Philip Howard from The Square restaurant, their own informal label which they place on their menus – a logo which states: 'Avoiding GM Foods' (BBC, 1999). It is interesting to note that both Barbaro and Carluccio are living abroad and promote the values of their regional cuisine within a globalised world.

Another example of chef activism against GM food is Tom Colicchio. In March 2016, the American celebrity chef who is of Italian descent presented to the US Senate a petition signed by 4,053 American chefs, asking it to reject a proposed bill that argues for voluntary labelling of GMO food. The petition was written by an NGO, Food Policy Action, which was co-founded by Colicchio (Bravo TV, 2016).

Looking more closely at Italy, it is clear that Italians have been active in and supportive of the alterglobalist movement. For example, many citizens took part in a protest against the annual G8 summit. The Genoa protest was, in fact, formative in the global justice and solidarity movements (Osterweil, 2013, p. 33). In trying to answer the question as to why Italy is so involved in 'a global movement ostensibly aimed at opposing neoliberalism and corporate-driven globalization', Osterweil argued that Italy has a particular history and political culture with which the global alterglobalist movement resonates (Osterweil, 2013, p. 33). Food is definitely part of this culture.

Another alterglobalist movement that has built upon Italian culinary values is Slow Food, which comprises of a network of organisations called Slow Food in Europe, the US and Japan. Today the network unites 100,000 members who are of diverse backgrounds: 'chefs, youth, activists, farmers, fishers, experts and academics' who are 'passionate about good, clean and fair food' (Slow Food, 2015a). The movement was created after a demonstration at the intended site of a McDonalds at the Spanish Steps in Rome, thus a comparison with Bove's campaign is unavoidable. The organisation was officially registered in Paris when the Slow

Food Manifesto was signed in 1989 (Slow Food, 2015b). Carlo Petrini, a co-founder of the movement, explained that the Italian activism was motivated by a strong cultural commitment, widespread among smallholders, and featured a tendency to form organised groups, called associazionismo (Petrini, 2003, pp. 1–3). The first civic associations that focused on the cultural value of Italian heritage in agriculture and food were the Free and Praiseworthy Association of the friends of Barolo, a famous Italian wine, ARCI (Associazione Ricreativa e Culturale Italiana) and Arcigola, formed in the 1980s. They aimed 'to create awareness of local products and awaken people's attention to food and wine and the right way to enjoy them'. They were quite innovative in their attempts to promote local agricultural products, first through tasting courses, get-together events and mail order sales, starting a cooperative movement (Petrini, 2003, pp. 4–5).

Members of the boards of these civic organisations formed the Slow Food Movement which merged with Arcigola. In the 1990s, Slow Food launched its publishing logo, and it became 'a powerful vehicle of identity' which allowed it to counteract the exclusive focus of media on high end restaurants and provide a chance for modest Italian eating establishments (Petrini, 2003, pp. 7–8). The movement's ideology was to use the notion of the taste of food and social justice values, combining ecology, support of regional producers, gastronomy and civic activism (Petrini, 2003, pp. 8–12). The common identity was built upon four major themes: sharing knowledge of food production; cherishing the artistic, historical and environmental values of food; protection of consumer and producer; and promotion of 'the pleasures of gastronomy' (Petrini, 2003, pp. 12–13).

Following these values, Slow Food Europe developed its stand against GM crops on the basis of seven arguments. It argued that GM crops leave 'biodiversity at risk', create 'toxic crops, toxic land', and signify 'corporate control on [the] rise', leaving 'small-scale farmers under threat', with 'no space for food culture', disseminating 'poor facts on health and safety', along with the 'the myth of feeding the world' (Slow Food Europe, 2016). Since then, Slow Food International has adopted the same views and has referred to ISAAA, Friends of the Earth and GMWatch as its information sources (Slow Food, 2015b).

The movement began in Italy and retains strong links there but also quickly spread across Europe and reached the US in 2000 (Slow Food, 2015b). It is worth mentioning that Slow Food USA, in the country that is most arguably favourable to the use of GM crops, shares opposition to GM with other organisations in the network. Several articles and blogs have been written educating consumers about the risks of GM food, and it has adopted the anti-corporate rhetoric criticising Monsanto, supporting smaller groups running campaigns against GM crops, such as 'No on 37: Coalition Against the Deceptive Food Labelling Scheme' (Yowell, 2013).

Thus, in this case, the values of taste and regional identities were incorporated into the alterglobalist rhetoric, successfully transplanted and used in different institutional and cultural environments. We can explain the popularity of Slow Food values in Italy and Europe as generally occurring where a regional culinary capital of a European country can be a tool to defend the interests of local producers. In the case of overseas partners within the network, it is the values of gastronomy and social and environmental justice that have a much broader connotation serving the common ground for movement members, while also including the same regional culinary capital. It is indeed interesting how regional culinary capital has been used as a mobilisation tool to run different campaigns, particularly against GM crops in different Western countries. While in Italy, the concept of territory was well defined in geographic terms – regional and local – in the US, the Italian culinary capital referred more to food safety and taste and passed over geographic terms.

Alterglobalist campaigns against Monsanto

Quite often, the object of protests from the alterglobalist movement are multinational corporations or MNCs; these are seen to embody all the negative aspects of globalisation. A big business corporation is often cursed for its capacity to combine and execute power over a vast number of people and promote social and economic inequality, thus injustice. Even in 1933, corporations were referred to as 'Frankenstein monsters' in court (Bakan, 2005, p. 8).

Among those MNCs attacked by activists, Monsanto is presented as the most evil: 'a calmly arrogant company heedlessly profiting from suffering of victims and the destruction of ecosystems' (Hulot, 2010, p. IX). It would be interesting to count the exact number of anti-Monsanto campaigns and associated organisations. The exact number is unknown, but it would likely reach a few dozen. These include small campaigns within the US where the company bases its headquarters, such as the march of African Americans in Anniston, Alabama, against Monsanto, blaming the company for the loss of relatives to cancer (Robin, 2010, p. 11), and large global campaigns.

The Organic Consumers Association, an American NGO that campaigns 'for health, justice, sustainability, peace and democracy' and is based in the state of Minnesota, began its anti-Monsanto campaigns in the 1990s. Today, it uses its website to fundraise for campaigns, files suits against Monsanto and lobbies authorities against the voluntary labelling of GM crops. In addition, it arranges public events such as the Consumer Revolution Campaign, People's Assembly, March against Monsanto, and the International Monsanto Tribunal to be held in The Hague in 2017 (OCA, 2016a). Both the Tribunal and the Assembly are held in The Hague, known as 'the international city of peace and justice'. These are simulated tribunals against Monsanto where a panel of international judges and lawyers will hear testimony from victims of the company's products. The end of the tribunal is marked by a march called the 'Global World Food Day with March against Monsanto', the main aim of which is to raise awareness about 'Monsanto's crimes against nature and humanity' (OCA, 2016b). The issues to be raised at the tribunal are the right to a healthy environment, right to food, right to health, freedom of expression and academic research and complicity in war crime and ecocide. The Tribunal steering committee includes activists and academics who have publicly spoken out against GM crops: Vandana Shiva, Corinne Lepage, Marie-Monique Robin, Olivier de Schutter, Gilles-Éric Séralini, Hans Herren, Arnaud Apoteker, Valérie Cabanes, Ronnie Cummins and Andre Leu (OCA, 2016b). Monsanto is chosen as a target because it is symbolic of 'impunity of transnational corporations and their management' and 'represents the worst of the worst when it comes to buying political influence and inflicting harm' (OCA, 2016b). The funding for the event was raised by the Organic Consumers Association and Regeneration International through individuals and other NGOs (OCA, 2016b).

Another global campaign is named March against Monsanto and was run for the first time in 2012 by the Occupy movement, another branch of the alter-globalist movement. In 2015, the march was held simultaneously on May 24th in around 400 cities in more than 40 countries. Each march had a specific focus although promoted as being part of a global campaign. For example, in the capital of Burkina Faso, Ouagadougou, 500 people marched in protest against GM cotton introduced by Monsanto (Agence France-Presse, 2015).

In Cambridge, where I live, an annual march against Monsanto is held. In 2015, the march was presented by three people sitting outside King's College on the King's Parade, the heart of the academic town. Upon asking them about their motivation, they told me that Monsanto corrupts science and destroys fair science and commented that they were proud to be connected to groups in other countries. Appendix 2 provides their manifesto for the year 2015.

Literature

Agence France-Presse (2015) Tens of thousands march worldwide against Monsanto and GM crops. *The Guardian*, 24 May 2015. www.theguardian.com/environment/2015/may/24/tens-of-thousands-march-worldwide-against-monsanto-and-gm-crops as viewed 25.08.2016.

Aries, P., Terras, C. (2000) *Jose Bove la revolte d'un paysant*. Villeurbanne: Editions Golias.

Asia Institute (2003) Philippines' Prof. Walden Bello wins Right Livelihood 'Alternative Nobel Prize' for 2003. Published online 8 October 2003. http://web.international.ucla.edu/asia/article/5008 as viewed 25.08.2016.

Bakan, J. (2005) *The corporation: The pathological pursuit of profit and power*. London: Constable.

Barbaro, G. (2013) *Family, food and friars: Experience the richness of Italian cuisine through cultivating, cutting, cooking and consuming with those you love*. New York: Ginjules Publishing.

BBC (1999) UK chefs take GM food off the menu. *BBC News*, 21 April 1999. http://news.bbc.co.uk/1/hi/uk/324592.stm as viewed 25.08.2016.

Belasco, W. (1989) *Appetite for change: How the counterculture took on the food industry, 1966–1988*. New York: Pantheon.

Belasco, W. (2008) *Food: Key concepts*. Oxford: Berg.

Bello, W. (2000) From Melbourne to Prague: The struggle for a deglobalized world. Talk delivered at a series of engagements on the occasion of demonstrations against the World Economic Forum (Davos), Melbourne, Australia. 6–10 September 2000. https://ratical.org/co-globalize/WB0900.html as viewed 25.08.2016.

Bello, W. (2004) *Deglobalization: Ideas for a new world economy*. London: Zed Book.

Bello, W. (2009) *The food wars.* London: Verso.

Best, S. (2011) Introduction, in S. Best (ed.) *The global industrial complex: Systems of domination.* Lanham: Lexington Books, pp. ix–xxv.

Bleiker, R. (2002) Politics after Seattle: Dilemmas of the anti-globalization movement. *Pacifica Review,* 14(3): 191–207.

Bove, J. (2007) *Candidat Rebelle.* Sarthe: Hachette Literatures.

Bove, J. (2015) Speech in plenary: Use of genetically modified food and feed. P8_CRE-REV(2015)10–28(4). Strasbourg, 28 October 2015. www.europarl.europa.eu/sides/getDoc.do?pubRef=-//EP//TEXT+CRE+20151028+ITEM004+DOC+XML+V0//EN&language=en&query=INTERV&detail=3-035-000 as viewed 25.08.2016.

Bove, J., Dufour, F. (2001) *The world is not for sale: Farmers against junk food.* Interviewed by Gilles Luneau. London: Verso.

Bravo TV (2016) Tom Colicchio and 4000 chefs call for urgent action on food law. Published online 10 March 2016. www.bravotv.com/blogs/tom-colicchio-4000-chefs-petition-senate-for-action-on-GMO-law as viewed 25.08.2016.

Calhoun, C., Sennett, R. (2007) *Practicing cultures.* New York: Routledge.

Carluccio, A. (2001) Interview with Antonio Carluccio: Back to good will hunting Antonio Carluccio talks to Catherine Brown and reveals the real magic of mushrooms as well as the benefits of foraging for free. *The Herlad,* 27 October 2001. www.heraldscotland.com/news/12135129.Back_to_good_will_hunting_Antonio_Carluccio_talks_to_Catherine_Brown_and_reveals_the_real_magic_of_mushrooms_as_well_as_the_benefits_of_foraging_for_free/

Carluccio, A. (2012) *Simple cooking.* London: Quadrille Publishing.

Confédération Paysanne (2016) Qui Sommes Nous. www.confederationpaysanne.fr/gen_article.php?id=8&t=QUI%20SOMMES-NoUS%20? as viewed 25.08.2016.

Crawford, S. (2005) The protests of the Anti-Globalization Movement. http://web.uvic.ca/~stucraw/Lethbridge/MyArticles/AntiGlobalization.htm as viewed 26.08.2016.

Focus (2016a) Who we are. http://focusweb.org/about

Focus (2016b) Partners. http://focusweb.org/content/alliances-networks

Held, D., McGrew, A., Goldblatt, D., Perraton, J. (1999) What is globalization? www.polity.co.uk/global/whatisglobalization.asp as viewed 26.08.2016.

Heller, C. (2012) *From food, farms and solidarity.* Durham: Duke University Press.

Hulot, N. (2010) A book for public health, in M.M. Robin (ed.) *The world according to Monsanto: Pollution, corruption, and the control of the world's food supply.* New York: New Press, pp. IX–XII.

Johnston, J., Baumann, S. (2010) *Foodies: Democracy and distinction in the gourmet foodscape.* New York: Routledge.

Juris, J. (2008) *Networking futures: The movement against corporate globalization.* Durham: Duke University Press.

Kaci, R., Kerhuon, Y. (2007) *Qui se cashe derrière les moustaches de Jose Bove?* Paris: Editions des Syrtes.

Kahn, R., Kellner, D. (2007) Resisting globalization, in G. Ritzer (ed.) *The Blackwell companion to globalization.* Malden: Blackwell Publishing, pp. 662–674.

Klein, N. (2004) Reclaiming the commons, in T. Mertes (ed.) *A movement of movements is another world really possible?* London: Verso, pp. 219–229.

McMichael, P. (2007) Globalization and the Agrarian world, in G. Ritzer (ed.) *The Blackwell companion to globalization.* Malden: Blackwell Publishing, pp. 216–238.

Morena, E. (2013) Constructing a new collective identity for the alterglobalization movement: The French Confédération Paysanne (CP) as anti-capitalist 'peasant' movement, in C. Fletcher Fominaya, L. Cox (eds.) *Understanding European movements: New social movements, social justice struggles, anti-austerity protests.* Abingdon: Routledge, pp. 94–105.

Naccarato, P., LeBesco, K. (2012) *Culinary capital.* London: Berg.

Organic Consumers Association (OCA) (2016a) Millions against Monsanto: About the campaign. www.organicconsumers.org/campaigns/millions-against-monsanto as viewed 26.08.2016.

Organic Consumers Association (OCA) (2016b) International Monsanto Tribunal frequently asked questions. www.organicconsumers.org/news/international-monsanto-tribunal-frequently-asked-questions as viewed 25.08.2016.

Osterweil, M. (2013) The Italian anomaly: Place and history in the Global Justice Movement, in C. Fletcher Fominaya, L. Cox (eds.) *Understanding European movements: New social movements, social justice struggles, anti-austerity protests.* Routledge: Abingdon, pp. 33–46.

Petrini, C. (2003) *Slow food: The case for taste.* New York: Columbia University Press.

Pingaud, D. (2002) *La Longue Marche de Jose Bove.* Paris: Editions du Seuil.

Robin, M.M. (2010) *The world according to Monsanto: Pollution, corruption, and the control of the world's food supply.* New York: New Press.

Shiva, V. (1999) This round to the citizens. *The Guardian*, 8 October 1999. www.theguardian.com/society/1999/dec/08/wto.guardiansocietysupplement1

Shiva, V. (2011) The agricultural industrial complex, in S. Best (ed.) *The global industrial complex: Systems of domination.* Lanham: Lexington Books, pp. 169–196.

Slow Food (2015a) Our network. www.slowfood.com/our-network/ as viewed 25.08.2016.

Slow Food (2015b) Our history. www.slowfood.com/about-us/our-history/ as viewed 25.08.2016.

Slow Food Europe (2016) GMO. www.slowfood.com/sloweurope/en/topics/ogm/ as viewed 25.08.2016.

Spooner, B. (2015) Globalization via world urbanization: The crucial phase, in B. Spooner (ed.) *Globalization: The crucial phase.* Philadelphia: University of Pennsylvania Museum of Archaeology and Anthropology, pp. 1–21.

Terral, P.M. (2011) *Larzac De la Lutte Paysanne a L'Altermondialisme.* Toulouse: Editions privat.

Trouillot, M.R. (2003) *Global transformations.* New York: Palgrave MacMillan.

van der Ploeg, J. (2008) *The new peasantries.* London: Earthscan.

Waters, S. (2012) *Between republic and market: Globalization and identity in contemporary France.* London: Continuum.

Watson-Smyth, K. (1999) Ban 'freak' modified foods, say top chefs. *Independent*, 27 January 1999. www.independent.co.uk/news/ban-freak-modified-foods-say-top-chefs-1076509.html as viewed 25.08.2016.

Yowell, E. (2013) Know your GMOs: The 'Monsanto rider'. www.slowfoodusa.org/blog-post/know-your-gmos-the-monsanto-rider as viewed 25.08.2016.

7 The feminist discourse

Introduction to the discourse

Eco-feminism derives from feminism, which is the theory of political, economic and social gender equality. Koppel defined it as 'a social, political, and academic movement that views the oppression of women and the exploitation of nature as being interconnected' (2011, p. 119). It is also an area of academic study, although it differs from traditional Gender Studies.

It came as 'a new term for an ancient wisdom' in the 1970s, deriving from the reference to different social protest movements (Mies and Shiva, 1993, p. 13). To be exact, the first to use the term was Francoise d'Eaubonne, a French militant radical feminist, who took part in the Resistance and co-founded the Women's Liberation Movement (Mouvement de liberation des femmes) (Roth-Johnson, 2013, p. 52). She also founded the Ecology and Feminism Movement (Mouvement Ecologie et Feminisme) in 1978 which combined two topics – female emancipation and environmental conservation. The term 'eco-feminism' appeared for the first time in her book *Feminism or Death* in 1974, was developed in later writing and soon adopted by the feminist movement. D'Eaubonne argued that the 'ecological drama directly results from the patriarchal system'. This system relies on the 'appropriation of agriculture and discovery of fatherhood by men' ending the previous epoch, before neolith, when agriculture and fertility was owned by women (d'Eaubonne, 1978, p. 23). She described the global processes of human impact on the environment, warning against a possible apocalypse and called for direct action – to cease the cause of such destruction, the patriarchal system (d'Eaubonne, 1978, p. 23). This call to overturn the male-dominated world suggests that eco-feminism is a social movement based on social utopia.

Her ideas were soon adopted by other feminists at the first eco-feminist conference in 1980 – 'Women and Life on Earth: A Conference on Eco-feminism in the Eighties' – where 'connections between feminism, militarisation, healing and ecology were explored' (Mies and Shiva, 1993, p. 14). Industrial accidents in La Roche Givaudan, Switzerland, Bhopal, India, and Chernobyl in the USSR were named as catastrophes against women as they particularly affected the well-being and health of women, illustrating the main point of eco-feminism that women and nature are oppressed under the modern rule of patriarchy (Mies and Shiva, 1993, pp. 14–15).

Merchant (1995) argued that eco-feminism has elements of many types of feminism: liberal, Marxist, third world, cultural, social and socialist and, thus, it is a diverse movement. In general, eco-feminism is a social movement that discusses the connection between women and nature, but it also connects other movements, such as those against racism and social inequality. In the words of its ideological leaders, feminist research is

> a criticism of the prevailing paradigm of science and social science, which had not only made women and their contribution invisible, but was most profoundly imbued with androcentric, that is, male-centered prejudices, both in its general assumptions and conceptualizations and in its theories and methods.
>
> (Mies and Shiva, 1993, p. 37)

To summarise, the general principles of eco-feminism include:

- The domination of women and the domination of nature (as well as other forms of domination such as racism, sexism, and social inequality) are interconnected.
- This is justified by a hierarchy that eco-feminists seek to resist on all levels.
- Dualistic thinking, in particular the distinction between culture and nature, supports this domination.
- The central goal is to replace these dominative policies, practices, and philosophies with ones that are not.

(Koppel, 2011, p. 119)

Eco-feminism represents ideas of radical or deep ecology, an ideology developed around the same time which generally represents a biocentric rather than anthropocentric worldview. The early examples of eco-feminism activity are protests and campaigns by women to save and restore the environment, such as the Green Belt Movement in Kenya and the Chipko Movement in India (Gnanadasan, 2000). One of the obvious leaders of modern eco-feminists is Dr Vandana Shiva. In 1993, she published, with Maria Mies, a book entitled *Ecofeminism*, which can be used as a primary source for understanding eco-feminist discourse within the GM crop debate. The union is not accidental, as Mies had written about Indian women, colonisation and patriarchy in the 1980s. Other ideological leaders of eco-feminism include Susan Griffin, Rosemary Radford Ruether, Carolyn Merchant, Irene Diamond, Ynestra King and Ariel Salleh.

Eco-feminism as an alternative ideology is based on the radical critique of modern society. In its turn, criticism of eco-feminism consists of 'essentialism claims', that is, claims that 'cross-culturally and cross-historically, all people of a particular gender or other category share the same traits' (Koppel, 2011, p. 121). Women who do not share eco-feminist values are called passive, as they are co-opted by the luxuries of Western life (Mies and Shiva, 1993, 1993; Shiva, 2002).

Eco-feminism in some instances promotes the notion of superiority – 'women are perceived as having an epistemologically privileged understanding of nature, whereas men are seen as being more related to culture', as though they are not connected to nature (Koppel, 2011, p. 121). This argument, however, is not present in Shiva's feminist ideology since she accepts that 'ontologically, there is no divide between man and nature, or man and woman, because life in all its form arises from feminine principle' (Shiva, 2002, p. 40).

It would be logical to assume that all eco-feminist writers and activists are women, yet, there are men who have also contributed to the development of ideas that are used in eco-feminism, such as James Lovelock. My interviewee, Muna Lakhani, calls himself a male feminist. Another point is that 'eco-feminism has been criticized for diverting the focus of the environmental movement into the feminist movement and for running the risk of oversimplification in suggesting that abolishing patriarchy would be the "magic solution" for the social and environmental crisis' (Koppel, 2011, p. 121). As there is arguably no quick fix, Koppel asked how eco-feminism could most efficiently contribute to appealing to people to reduce environmental degradation (Shiva, 2002, p. 40). This chapter examines the main postulates of the eco-feminist theory and reviews campaigns centred on women run by Navdanya, Vandana Shiva's NGO.

Gaia theory

Originally, Gaia was the name of the Greek goddess of Earth, becoming a metaphor for the description of the biological processes of the planet. The Gaia theory or hypothesis uses a 'medical metaphor of homeostasis' to describe the complexity of the Earth as a system (Duffy, 2011, p. 187).

The author of the theory is actually a man, Dr James Lovelock. A medic by training, he has become an independent scientist involved in revolutionary research, inventing 'an instrument that was so sensitive at detecting minute traces of atmospheric pollutants that many colleagues refused to believe his claims'. As a result, he 'ranged freely across the disciplinary boundaries that too often constrain "institutional" thinkers' and 'is beholden to no institution' (Reese, 2009). According to Lovelock, he became 'a radical scientist . . . also because the scientific community [was] reluctant to accept new theories'. In 1968, while working in a bioscience lab in the US, he started to develop his ideas about the Gaia Earth (Lovelock, 1995, p. xix). According to his theory, the earth is considered 'a living organism', so it is argued that the evolution of the species and natural selection should be looked at from a 'planetary perspective', the health of the planet and its inhabitants are connected, and human activity, particularly agriculture and forestry are the most serious source of damage (Lovelock, 1995, p. xix).

Intellectual inspiration came from a Russian scientist, Vladimir Vernadsky, his great uncle Evgraf Korolenko who was a philosopher, and James Hutton. Vernadsky developed a concept of the biosphere or Earth's global ecosystem, while Hutton also wrote about 'living Earth' in his studies of geology. Following these inspirations, Lovelock arrived to his conclusion that Gaia is a 'super organism'

(Lovelock, 1995, p. 15). Within this super organism, genetics play an important role, 'acting as a repeater' through 'transmitted coding messages' (Lovelock, 1995, p. 64). It is a precise channel of information to carry on life as it is, as a self-regulating system, providing stability and precise control mechanisms. The choice for the name of his theory came through his friend, Bill Golding, who had a better grounding in Classics and suggested Gaia, the Greek goddess (Lovelock, 2000c). While in the US, Lovelock got in contact with Lynn Margulis, who read science in the University of Chicago and specialised in zoology and genetics. She added to the Gaia theory a concept of symbiosis as 'the apparently cooperative relationship between two organisms from which both benefit' (Turney, 2003, p. 34). Eventually, she became fascinated by Lovelock's hypothesis, introducing to it the endosymbiotic theory. Upon meeting, they discussed and shared their ideas about biota search as a system. They published two papers which went fairly unnoticed and only the first edition of *Gaia: A New Look on Life on Earth*, published in 1960 by the Oxford University Press, revealed their ideas to the public (Turney, 2003, p. 48). The book suggested that the Earth was more than a sum of its parts:

> If Gaia does exist, then we may find ourselves and all other living things to be parts and partners of a vast being who in her entirety has the power to maintain our planet as a fit and comfortable habitat for life.
>
> (Lovelock, 2000a, p. 1)

In all his works, Lovelock stressed that Gaia was a scientific hypothesis (Turney, 2003, p. 53). The theory was not considered by most scientists to be science and some of its postulates are challenged. Vedwan, for example, suggested that Lovelock, in an attempt to win over critique of the unscientific ground of his research, made a reference to Darwinian evolution. Lovelock argued that the environment modifies to create favourable conditions to organisms, but not everyone is convinced that there is intention to support life. Arguments that are put forward to contradict the Gaia theory refer to mass extinction events (Vedwan, 2011, p. 240). Richard Dawkins dismissed the Gaia theory:

> The fatal flaw in Lovelock's hypothesis would have instantly occurred to him if he had wondered about the level of natural selection process which innovation in the modern synthesis was however the new conception that a population that was deemed to undergo evolution could best be thought of as a population of fundamental replicating units – of genes – rather than as a population of individual animals or of cells.
>
> (Dawkins, 1999, pp. 238–239)

The gender factor played its role in the theory's acceptance. 'No feminist will complain if I use the pronoun, she, for Gaia, but no [scientist] journal will publish my paper if I do', wrote Lovelock (Lovelock, 1995, p. xiii). He is aware that his concept has been welcomed and employed by the green movement, particularly because of its symbolism (Lovelock, 1995, p. xiv). Yet he himself did not join the environmental movement or its eco-feminist aspect. Despite the fact that he knew Rachel Carson and discussed with her the damage to the ecosystem, he has become a critic of the ecological movement:

> We needed common sense and the acceptance of the wisdom of the physician Paracelsus who said long ago, 'The poison is the dose'. . . . Unfortunately, common sense is a rare commodity. . . . Greens are not just ignorant of science; they hate science.
>
> (Lovelock, 2000b, p. 200)

Despite attempts by Lovelock and Margulis to 'disavow themselves from the theological interpretations, a cult of Gaia is emerging and blending with ecofeminism and neopaganism' (Werner, 2011).

As a result, Lovelock is not a member of any of the NGOs that form the Gaia movement, such as the Gaia Foundation, Foundation for Gaia, Project of Gaia, Gaia-Oasis and the Gaia Institute to name a few. Lovelock's ideas on GM crops are different from those of the Gaia Foundation, a well-known opponent of GM crops: 'My feeling about GM crops is that we are straining at a gnat while swallowing the camel of greenhouse-gas accumulation' (Lovelock, 2000c).

Eco-feminist ideas on women and biotechnology

Position of women in modern society

Most eco-feminists are defining the role of women in society, which is subordinated and discuss further search for identity and diverse models which could better accommodate women and their special role of nature carers in modern society. Modern society is dualistic and subordinated: 'Nature is subordinated to man; woman to man; consumption to production; and the local to the global, and so on' (Mies and Shiva, 1993, p. 5). Patriarchal rule brings inequality north and south, colonisation of women, nature and culture are in an antagonistic relationship (Mies and Shiva, 1993, p. 5). This same patriarchy approves of cultural relativism, which accepts violent practices and the exploitation of women such as through the payment of a dowry and the practice of female genital mutilation (Mies and Shiva, 1993, p. 11). It is argued that women as well as nature 'pass[es] through the same stages as the traditional sexual relationship: aggression, conquest, possession, control' (Mies and Shiva, 1993, p. 15).

As women fall into the Other category, as do people of colour and those of third world countries, they are the subject of colonisation (Mies and Shiva, 1993, p. 55). It is argued that they are an 'internal colony' of industrial society and that the 'patriarchal-capitalist sexual division of labour' does not take women or future generations into consideration, nor does it consider the ecological cost associated with the capitalist production (Mies and Shiva, 1993, p. 58). There is a general critique of industrial civilisation which is associated with poverty, criminality, depression and suicide, violence against women and children, urban emissions and fast food, all indicating lower life quality and deprivation of the basics such as 'clean air, pure water, healthy food, space, time and quiet' (Mies and Shiva, 1993, p. 61).

The colonisation of women is also understood as 'socio-psychological status', as the state and its family policy controls women, their sexuality, fertility and work capacity, 'housewifisation' is imposed upon women with the aim of unlimited economic growth (Mies and Shiva, 1993, pp. 119–120). Vandana Shiva used her argument against cash monocrops in favour of subsistence farming to illustrate 'how women are devalued' in the market economy; girls and women are treated marginally in food distribution and often suffer from malnourishment (Mies and Shiva, 1993, p. 73). However, emancipation for women and gender enlightenment 'as a catch up with the men' is considered to be the wrong strategy (Mies and Shiva, 1993, pp. 7–8). The same applies to modernisation: it does not bring equality but destroys the culture of non-modern societies and the planet's biodiversity (Mies and Shiva, 1993, p. 11). Even international political efforts to address environmental degradation, such as the Earth Summit of 1992, are only reflections of capitalist patriarchy (Mies and Shiva, 1993 p. 33). Such environmentalism becomes 'the rhetoric of managers-technocrats, who see the ecological crisis as an opportunity for new investment and profit'. It becomes a 'new means of dispossessing the poor' from global commons to preserve natural resources for the rich (Mies and Shiva, 1993, p. 86). Women through their fertility are transformed from victims into villains responsible for overpopulation (Mies and Shiva, 1993, p. 87). The 'feminist perspective' is argued to prevent such violence and injustice; as it goes beyond the categories of patriarchy such as power structure and the narrow definition of nature, it brings 'the context of regeneration' (Shiva and Mies and Shiva, 1993, p. 33).

Different from men, women are perceived as planetary caretakers and cleaners responsible for this regeneration and sustaining the life. Yet even in this there is an element of inequality. Men expect that 'somebody always comes along later to clean up. Like Mom' (Spretnak, 1987, p. 6). As a result, women carry a double burden: to protect themselves and nature (Merchant, 1995, p. xi). An example of such female responsibility is the Australian feminist campaign which lobbied the federal government in 1990 against GMOs in which activists ran a community course and wrote publications under the title of eco-feminist action (Merchant, 1995, p. 25).

Science, scientists and gender

Eco-feminists argue that science and technology are not gender neutral (Mies and Shiva, 1993, p. 3). Even disciplines within social science, such as Women's Studies, have become instruments for the oppression and exploitation of women (Shiva and Mies, 1993, p. 37). Eco-feminists complain that Western science is paternalistic, viewing women as objects and supporting the oppression of women and nature which are considered to be closely connected.

Schiebinger pointed out that traditionally – long before the Gaia hypothesis – from Aristotle, Darwin and Freud, nature was perceived as gender based and sexuality based (Schiebinger, 1993, p. 1). Botanists such as Carl Linnaeus studied the monogamy and polygamy of plants and introduced ideas of sexual difference and hierarchy (Schiebinger, 1993, p. 1). This led some socio-botanists to claim that 'rape' in the plant world was acceptable and justified the passive, dependent role of women in society by 'genetic origin'. This is sharply opposed by eco-feminists (Bleier, 1984, p. 46). Merchant (1996) proposed that femininity symbolised 'enclosure', with the female womb referring to an enclosed garden that keeps its secrets and bears fruit. Men are witnesses and voyeurs trying to understand how nature works under the terms of Western science (Merchant, 1996, p. 66). In modern times,

> new developments in biotechnology, genetic engineering and reproductive technology have made women acutely conscious of the gender bias of science and technology and that science's whole paradigm is characteristically patriarchal, anti-nature and colonial and aims to disposes women of their generative capacity as it does to the productive capacities of nature.
>
> (Mies and Shiva, 1993, p. 16)

Thus, there are two types of knowledge: specialist, 'reductionist', based on expertise and ignorance and common knowledge of which women are natural bearers.

In the modern science imposed by the West, women and nature become the object of knowledge, a 'raw material' to be studied (Mies and Shiva, 1993, p. 25). Shiva quoted Oakley, who referred to medical studies of female fertility wherein women are seen as mere containers, a resource for reproduction (Mies and Shiva, 1993, p. 27). 'Modern gynaecological profession, in vitro fertilisation, surrogacy, and biotechnology research mimic women's generative capacities, carrying men's power to great heights' (Merchant, 1996, p. 25).

Mies described Western scientists as 'new patriarchs' who do not recognise their responsibility or see themselves as citizens or fathers, have abstract ideals and would accumulate knowledge 'at any price' if they could (Mies and Shiva, 1993, p. 50). Thus, in her view, they would have to give up their research on genes and reproduction, as the ethics applied are reactive and tend to hide their interventions and fabrications (Mies and Shiva, 1993, p. 50). She also claimed that modern science is male-dominated and agreed with Brian Easlea that science is driven by men's sexual desires, such as the 'phallic birth fantasies of these fathers of atom bombs and rockets' (Mies and Shiva, 1993, p. 50). They also tried to uncover secrets of femininity and nature by 'seducing . . . female land' (Merchant, 1996).

Eco-feminists such as Merchant and Shiva have referred to a number of famous Western scientists as patriarchs suppressing nature. Included in this are Francis Bacon, Thomas Hobbes, Joseph Glanvill, Robert Boyle, and George Sarton (Merchant, 1996, p. 61). Francis Bacon, who is called a 'new father of this natural science,' is mentioned, as his 'imagery treats nature as a female to be tortured through mechanical inventions' and 'strongly suggests the interrogations of the witch trials and the mechanical devices used to torture witches' (Merchant, 2008).

Bacon's methodology is also discussed by other eco-feminists. He invented new empirical method and advised 'the "new heroes of natural science" to brush aside all the old taboos without a qualm and to expose them as superstitions'. It is pointed out that taboos were first violated in developing countries by Western man (Mies and Shiva, 1993; Merchant, 1996).

In agriculture, nature and its embodiment in seeds are treated in the same manner as women. Seeds are colonised, as is the female body, and plant fertility is also redesigned. Intervention in seed reproductive function serves the same purpose – to control a common resource and turn it a corporate tool. The regeneration of seeds and their biodiversity is supplanted from farms and forests into labs (Mies and Shiva, 1993, p. 29). Shiva likened the withdrawal of seeds from the natural cycle of plant reproduction as being a divorce. She claimed that such commodified seeds do not reproduce, are 'ecologically crippled' and cause genetic erosion since indigenous varieties are abandoned (Mies and Shiva, 1993, pp. 30–32).

Biotechnology, following economics, pushes the assumption that it is possible to overcome the slow and primitive nature of seed-renewal through the introduction of transgenic seeds (Mies and Shiva, 1993, p. 28). But such an approach brings about only ignorance in relation to nature and women and a reductionist, mechanical view of plants as 'machines' (Mies and Shiva, 1993, p. 29). Scientists are portrayed as 'empirical men in white coats [who] steal DNA from spiders webs', missing continuity with nature and with an assertion of right, power without responsibility (Merchant, 1995, p. 66). As a result, this responsibility to care for nature falls to women.

Mies argued that 'unlimited progress' as seen by men is a myth and, thus, 'confidence in the ruling men in politics and science is dangerous': scientists are portrayed as greedy and ready to do questionable research for big money (Mies and Shiva, 1993, p. 94). Genetic engineering research is referred to as value free, moral free and ethic free. The politicians and scientists who make decisions regarding the application of technologies have been described as lacking imagination and emotions, particularly empathy. The case of Chernobyl is used as a reminder of irresponsible science, a warning against future catastrophes and an illustration to prove the fallacy of male claims over limitless knowledge, omniscience and male omnipotence, as those who created Chernobyl could not restore life after the catastrophe (Mies and Shiva, 1993, pp. 94–95). Knowledge itself is considered a social construct created by powerful social groups 'to claim privileged access to reality', and such power is revealed through the rhetoric of language (Merchant, 1996, p. 64).

To counteract such patriarchal science, eco-feminists have called for a return to a holistic approach that accepts women and nature as the subject and not object of research. For their part, eco-feminists have opposed the separation of science and politics and culture from nature, instead calling for social activism which goes beyond academic discourse (Mies and Shiva, 1993, p. 43).

Spirituality and religious aspect of eco-feminism

Under patriarchy, not all cultures are seen as equal; a dogmatic ideology is imposed universally with acceptance of cultural relativism, as shown earlier (Mies and Shiva, 1993, pp. 6–7). As the natural symbiosis and interconnectedness of life are ignored by patriarchy and diversity of life and culture are perceived as divisive and threatening, eco-feminism has called for a 'new cosmology and a new anthropology, which recognises that life in nature is maintained by means of cooperation and mutual care and love . . . forge a holistic all-embracing lifestyle' (Mies and Shiva, 1993, p. 6).

The patriarchal model turns even religion against women; monotheistic religions such as Christianity, Islam and Judaism are argued to be hostile to women and nature. To counteract, eco-feminists somehow have tried to revive 'a goddess-based religion'. The whole concept of spirituality is defined as 'the Goddess'. They have overcome the traditional division of spirit from the Material by suggesting that 'although the spirit was female, it was not apart from the material world, but seen as the life-force in everything and in every human being' (Mies and Shiva, 1993, p. 17). Reference is made to Starhawk, a writer of neo-paganism who defined spirituality as identical to 'women's sensuality, their sexual energy, their most precious life force, which links them to each other, to other life forms and the elements' (Starhawk, 1982, cited in Mies and Siva, 1993, p. 17).

A frequent argument of Vandana Shiva about the depletion of soil as a result of modern intensive agriculture is also dressed in the terminology of spirituality. Soil is called a 'sacred mother', as it not only allows the raising of crops but also serves as a 'spiritual home for most cultures' (Mies and Shiva, 1993, pp. 100, 102). Its destruction carries another spiritual meaning – the link to ancestries and traditions is broken (Mies and Shiva, 1993, p. 101). Soil is taken as the condition for the regeneration of nature and society. Shiva pointed out that the 'desacralisation' of soil is another form of colonisation and cites Rifkin, another anti-GM activist, who suggested that the desacralisation process 'serves as a kind of psychic ritual by which human beings deaden their prey, preparing it for consumption' (Mies and Shiva, 1993).

In the 1960s, Lovelock came up with his Gaia hypothesis and in 1978, Charlene Spetnak published her book, *Lost Goddess of Early Greece*. Under this influence, feminists in the 1970s attempted to create 'a new earth-based form of spirituality rooted in ancient traditions that revered both earth and female deities' (Merchant, 1995, p. 3). But the Gaia hypothesis in its spiritual meaning was not fully accepted since the concept was seen as 'problematic': 'Its message carries cultural baggage that undercuts its inspirational power. If Gaia is a self-regulating homeostatic system, then "she" can correct problems caused by humans or even find humans expendable' (Merchant, 1995, p. 4).

The female gender of Earth was considered unquestionable, but there were other deities apart from Gaia that have been referred to by eco-feminists. Allusions were made to Christianity, specifically the garden of Eden: the Earth used to be virgin and as undeveloped as Eden, then came the fall with disorder and chaos, with man being the creator who brings rule and law and is the agency of transformation (Merchant, 1995, p. 32). Interestingly, this transformation of nature comes with biotechnology: GM crops – in this case, the Flavr Savr tomato, which was the first GM fruit – help to redesign nature so 'the salinated irrigated desert can continue to blossom as the rose'. Transgenic agriculture refers to the recovery of the Garden of Eden where 'fruits [will] ripen faster, have fewer seeds, need less water, require fewer pesticides, contain less saturated fat and have longer shelf lives' (Merchant, 1995, p. 50). But this engineering of nature comes too close to the power of God and the DNA spiral becomes the bible snake (Merchant, 1995, p. 51). Thus, the intended recovery through genetic engineering is called 'fake' and 'the End Drama' might be just a social utopia in the 'post-patriarchal' style. The two real elements to recover paradise are argued to be women and nature (Merchant, 1995, p. 53). The non-Christian, third world pantheon provides more deities matching the eco-feminist ideology. For example, Gnanadason (2000) referred to Indian cosmology, which is based on the connectedness of person and nature, bringing such duality in unity through the image of vegetation coming from the body of the goddess Devi-Mahatmyam and rituals to restore the force of soil and search deities.

Eco-feminism and science fiction

Science fiction literature is another domain in which debates on GM crops are present. Interestingly, these have also reflected some of the ideas put forward by eco-feminism. It is my opinion that this is not a coincidence. Eco-feminism, as shown earlier, reflects the empowerment of women and refers to their creative powers to heal the world and fight injustice. This is often seen as a social utopia, although its ideologists might hope that a radical turnover of patriarchy is possible.

I refer to two science fiction writers: an American by the name of Paolo Bacigalupi who wrote *The Windup Girl* (2010), and Sergey Tarmashev, a Russian who in 2015 published a two-part epic novel called *Nasledie* (translates from Russian as 'The Heritage'). Both are young and write what can be called bio-punk, examining issues relating to biotechnology. Both write within the genre of ecotopia with a critique of the modern use of natural resources. Both works have received a reasonable amount of attention of readers and critics.

Bacigalupi explained his role as a science fiction writer to go 'two or three steps down the road beyond what you can actually report', to step over the conservatism of scientists and scientific journalists and 'illustrate a feeling or experience so that people can say, 'Does that seem like we want to be going toward?'' (Otto, 2012, p. 10).

Both writers refer to the twenty-third century in the context of possible future human history. It is also interesting to compare that such science fiction fits well into the catastrophism of non-science fiction works by Lester Brown, Paul Ehrlich and the Club of Rome. These too use extrapolation, which is the process of estimating beyond the original observation range. In the best traditions of the neo-Malthusian school, the central issue is food insecurity. Both novels can be called eco-feminist, as their main protagonists are female. Emiko, a Japanese 'windup' girl, and Alena, a Russian journalist, know about secret seedbanks that can restore human civilisation after it is realised that transgenic agriculture is a dead-end.

I came to know about Bacigalupi through a lecture by Dr Larsson at the University of Cambridge in Fall 2015 (Larsson, 2015). The second book was introduced to me by a Russian NGO – NAGS, which is described in Chapter 4 of this book.

They gave me a reprint of Tarmashev's two-volume work. I received *Nasledie* with comments that the main protagonist in this book is based on a living person named Elena Sharoykina, who is a director of NAGS. In his introduction, Sergey Tarmashev explained that the first volume may have 'too heavy read', as he uses too many scientific terminology but recommended it be read 'with all serious-ness'. He argued that the facts he refers to are 'correct and are based on works by many modern scientists – geneticists, biologist, and virologists'. He informs the reader that this is a real state of affairs in the modern world and that if no action is taken, the grim forecast shown in the second half will become 'a black reality for our children'. He also expresses 'a deep appreciation to all Russian scientists, selflessly developing national science under petty funding, avoiding the tempta-tion to exchange their dignity and national security for a generous rain of grants and awards' (Tarmashev, 2015a, pp. 5–6).

In his second volume, Tarmashev responded to the critique that his novel has too much scientific terminology for light reading: 'I respond to them: this book you don't have to read. It will tiresome your tender brain not used to work even further'. His work was created for people who are 'thinking and analytical and not ignorant'. Further, he also explained that the whole book was written only for the presentation of the real facts explained in the first part. Tarmashev made an emotional claim: 'Reader, remember! Now this bunch of uncorrupted rebels who are not willing to exchange their shame for silver coins are all that stands between the future of your children and those events described in the book' (Tarmashev, 2015b, p. 6).

The first volume depicts Elena Sharoykina as the daughter of an oligarch living a glamourous life, driving an expensive car, attending beauty salons and making friends with celebrities. She then discovers a global conspiracy to poison the world with GMOs and begins to make allies with American, Indian and Chinese people to protect the world. She carries on fighting against agricultural corporations from the UN forum and predicts that humanity will kill itself with GMO food. In the year 2053, she is depicted as an old lady, head of the International Association of Genetic Safety, an obvious allegory. In the meantime, humanity has already begun to degenerate, and many human babies are born as monsters who suffer from constant pain. In the second volume, readers are shown what happens in the year of 2267 when humans live in concealed protected zones which artificially provide habitable conditions because the outside environment is fully degraded and is in the control of mutants. One character, Prof. Sinyzin, makes a discovery to find the treasure left by the Great Sharo, or Forecasting Sharo, who is a goddess formed from the canonised image of Elena Sharoykina. Nonetheless, the novel does not have a happy ending.

Activism of Navdanya on women

As shown earlier, eco-feminism as a movement since its early days has focused on activism, and such activism has included fighting against genetic engineering for reproductive health and agriculture. In 1984, feminists formed the Feminist

International Network of Resistance to Reproductive and Genetic and Engineering (FINRRAGE). It organised several events, including international congresses, and mobilised women from universities, churches and trade unions which were opposed to the idea of a surrogate motherhood agency and genetic engineering in humans, animals and plants (Mies and Shiva, 1993, p. 16). Currently, the movement seems to be silent, as most position papers and articles date back to the 1980s and early 1990s (FINRRAGE, 2017).

There exist new generations of eco-feminist organisations and campaigns. In this instance, Vandana Shiva is the most active eco-feminist. In 1990, she wrote 'Most Farmers in India are Women' (FAO report) and founded the gender unit at the International Centre for Integrated Mountain Development (ICIMOD) in Kathmandu, Nepal. She also acted as a founder for the Women Environment and Development Organization (WEDO) and initiated another women's movement to work more specifically around the subjects of food, agriculture and biotechnology. This was called Diverse Women for Diversity (DWD) and was officially launched in Bratislava, Slovakia, in May 1998 (Chopra Foundation, 2016). The organisation's campaigns cover lobbying through public presentations, conferences and the issue of statements. In 2005, 75 women from 15 nations were present at DWD's conference (Navdanya, 2016a).

Navdanya's website introduces the 'global campaign of women on biodiversity and food security'. Other founders include Dr Jean Grossholtz, Professor Emeritus in Politics and Women Studies, Shiva's co-author, Beth Burrows, and German feminist, Dr Christine von Weizsäcker, who is president of a German NGO named Women in Europe for a Common Future (WECF) (Navdanya, 2016a).

DWD is an international platform for the global eco-feminist movement. At national and local levels in India, it works as the National Alliance of Women's Food Rights and Mahila Anna Swaraj (Women's Food Sovereignty). An example of its first campaigns is one against soybean-based food. In 1998, the National Alliance of Women's Food Rights, under the leadership of Dr Shiva, began a campaign against 'globalization of soybean-based food' which it argued was dangerous for female health and reproduction. It was in favour of traditional mustard oil. The organisation mobilised women from the slums of Delhi and challenged the ban on small-scale processing and sales of cooking oil in the Supreme Court of India (Shiva, 2000, p. 32).

The alliance has jointly with another 33 NGOs, including Greenpeace and women's organisations such as the All India Democratic Women's Association, signed the Women's Charter on Food Rights. The Charter has eight subchapters on the Indian constitution, nutrition, agriculture, public distribution, food prices, food exports, globalisation and genetic engineering in food crops (The National Alliance, date unknown). In 2001, DWD issued a statement on terrorism which included naming monocultures, biotechnologies and environmental degradation, alongside the US-led NATO policies as 'tools of terror'. Women and children were named as 'the worst victims of this reign of terror'. DWD condemned all acts of war and 'pledged to overcome this capitalistic patriarchal terrorism' (DWD, 2001).

Another set of activities centred on women that is run by Navdanya is the so-called Grandmother University. It was started in 2007 as a programme of Bija Vidyapeeth (the learning centre of Navdanya) and a continuity of tradition set by the villagers of Pattuvam in Kerala in 1997. It is a meeting point for older women ('grandmothers') passing their traditional knowledge on to younger generations (Navdanya, 2016b). The first course was launched in February 2007 and lasted three days. It contained presentations by Usha Maira, a great-grandmother who spoke about the importance of manual labour, singer Vidya Rao, activist Sudha Soni, Vandana Shiva's sister Dr Mira Shiva and community workers Shashi Tyagi and Rashmi Sharma (Goburdhun, 2007).

Literature

Bacigalupi, P. (2010) *The windup girl*. St. Ives: Orbit.

Bleier, R. (1984) *Science and gender: A critique of biology and its theories on women*. New York: Pergamon Press.

Chopra Foundation (2016) Speakers. Vandana Shiva. www.choprafoundation.org/speakers/vandana-shiva/ as viewed 01.08.2016.

Dawkins, R. (1999) *The extended phenotype: The long reach of the gene*. Oxford: Oxford University Press.

D'Eaubonne, F. (1978) *Ecologie, feminisme: revolution ou mutation?* Paris: ATP.

Duffy, L. (2011) Gaia hypothesis, in D. Mulvaney, P. Robbins (eds.) *Green politics*. Los Angeles: Sage, pp. 187–189.

DWD (2001) Statement. 1 October 2001. www.navdanya.org/archives/22-statement as viewed 01.08.2016.

FINRRAGE (2017) Genetic engineering. www.finrrage.org/?page_id=225 as viewed 04.05.2017.

Gnanadason, A. (2000) Towards a feminist eco-theology for India, in R.R. Ruether (ed.) *Women healing earth: Third world women on ecology, feminism and religion*. New York: Orbis Books, pp. 74–81.

Goburdhun, M. (2007) Grandmother's University promoting women's wisdom and traditional knowledge. *BIJA*, Spring–Summer 2007.

Koppel, M. (2011) Ecofeminism, in D. Mulvaney, P. Robbins (eds.) *Green politics*. Los Angeles: Sage, pp. 119–121.

Larsson, T. (2015) Seed and nation: Understanding divergent political responses to GMOs in Southeast Asia. Centre of South Asian Studies Seminar. University of Cambridge 25 November 2007.

Lovelock, J. (1995) *The ages of Gaia: A biography of our living earth*. Oxford: Oxford University Press.

Lovelock, J. (2000a) *Gaia: A new look at life on earth*. Oxford: Oxford University Press.

Lovelock, J. (2000b) *Homage to Gaia: The life of an independent scientist*. Oxford: Oxford University Press.

Lovelock, J. (2000c) Live chat. *The Guardian*, 29 September 2000. www.theguardian.com/books/2000/sep/29/scienceandnature.livechats as viewed 30.07.2016.

Merchant, C. (1995) *Earthcare: Women and the environment*. London: Routledge.

Merchant, C. (2008) 'The violence of impediments': Francis Bacon and the origins of experimentation. *ISIS*, 99: 731–760.

Mies, M., Shiva, V. (1993) *Ecofeminism*. London: Zed Books.

The National Alliance The National Alliance of Women's Food Rights. *BIJA*, 23/24.

Navdanya (2016a) Diverse women for diversity. http://navdanya.org/diverse-women-for-diversity as viewed 01.08.2016.

Navdanya (2016b) Grand Mother's University. http://navdanya.org/diverse-women-for-divesity/grandmothers-university as viewed 01.08.2016.

Otto, E.C. (2012) *Green speculations: Science fiction and transformative environmentalism.* Columbus: Ohio University Press.

Reese, M. (2009) Foreword, in J. Lovelock (ed.) *The vanishing face of Gaia: A final warning.* New York: Basic Books, pp. xiii–xvi.

Roth-Johnson, D. (2013) Back to the future: Françoise d'Eaubonne, ecofeminism and ecological crisis. *The International Journal of Literary Humanities*, 10(3): 51–61.

Schiebinger, L. (1993) *Nature' body.* Boston: Beacon Press.

Shiva, V. (2000) *Stolen harvest: The hijacking of the global food supply.* London: Zed Books.

Shiva, V. (2002) *Staying alive: Women, ecology and development.* London: Zed Books.

Spretnak, C. (1987) Ecofeminism: Our roots and flowering. *Ecoprint*, 3(2): 2–9.

Starhawk (1982). *Dreaming the dark: Magic, sex, and politics.* Boston: Beacon Press.

Tarmashev, S. (2015a) *Nasledie.* Moscow: AST.

Tarmashev, S. (2015b) *Nasledie 2.* Moscow: AST.

Turney, J. (2003) *Lovelock and Gaia: Signs of life.* Cambridge: Icon.

Vedwan, N. (2011) Gaia hypothesis, in H.S. Schiffman, P. Robbins (eds.) *Green issues and debates.* London: Sage, pp. 238–242.

Werner, M. (2011) Ecofeminism, neopaganism, and the Gaia Movement in the Postmodern Age. The Utopian Research. www.scribd.com/document/59253239/Michael-Werner-Ecofeminism-Neopaganism-and-the-Gaia-Movement-in-the-Post-Modern-Age?ad_group=Online+Tracking+Link&campaign=Skimbit%2C+Ltd.&content=10079&irgwc=1&keyword=4417&medium=affiliate&source=impactradius as viewed 30.07.2016.

Part 2

Processing the debate

In this part, we will discuss how the debate on GM crops is being processed. There are two arguments raised by the GM debate – the revision of the nature and creation of scientific knowledge and the political communication of public debates.

Political communication is constructed through messages and information produced by different political actors, such as the media, policymakers, NGOs and so forth. Within this model, citizens act as recipients (Entman and Bennett, 2001). In the democratic political process, there exists a match between the supply side of producing knowledge and demand side of communication, when citizens react to the information and knowledge offered by accepting it or being opposed.

The notion of public knowledge which is produced, communicated and responded to in public debate has developed from the framework offered by Jürgen Habermas. Public knowledge, and public opinion in relation to it, are developed in the public sphere.

Habermas defined the public sphere as 'opened to all': the public domain is opposed to private affairs, with an attribute such as public acknowledgement recognition or 'publicity' (Habermas, 2005, pp. 1–20). It is part of civil society and also characterised by representation. Different groups, often based on religion or class attribution, began to use the public sphere to declare their representation and influence public opinion. In medieval times, authorities and the ruling classes began to use the media as a tool in the public sphere to disseminate information and 'facilitate communication' within society in an aim to inform and educate; in other words, to share knowledge (Habermas, 2005, p. 24). This ensured that civic groups were also able to oppose the authorities within the public sphere, respond to official opinion with their own viewpoint, and present their reasoning for public use (Habermas, 2005, p. 27). So it became a question of power wherein the exchange of information is used for the legitimisation and domination of certain social and political groups.

The public sphere is a network for communicating information (it is reproduced via 'communicative action') and brings to the fore different points of view. To reach a common understanding, discourses and bargaining become 'problem solving forces' (Habermas, 2004, p. 326).

It is then possible to connect communicative networks with the production of different discourses. Yet Emden and Midgley (2013) argued that 'much of the debate about the role of the public sphere in political theory often proceeds without paying attention to the many contexts and discourses within which publics come into existence'. For that, they remind us that

> Publics are historically conditioned and socially concrete 'things'; that is, they emerge or are dissolved in specific locations, they are shaped by the technologies and material culture to which they are subject, and they also serve the production of knowledge.
>
> (Emden and Midgley, 2013, p. 3)

As we have seen in Chapters 2–7, different organisations, depending on their geography and major area of activity, shape different discourses in the debate on GM crops.

Thus, public knowledge is 'knowledge that circulates among specific constituencies': 'Knowledge about the world we live in – even if such knowledge is of a highly specialized nature and accessible, initially at least, only to a small group of people of experts' – then tends to become public knowledge (Emden and Midgley, 2013, p. 3).

Habermas proposed that two principles within the public sphere should be built: transparency and that 'only the force of the better argument should prevail' (Emden and Midgley, 2013, p. 4). That is the situation in the ideal world; however, in the real world, knowledge is produced and co-opted in a complex system of policymaking where specialised knowledge is reassessed and then transformed into public knowledge. For example, scientific knowledge 'is extended outside the laboratory not by generalization into universal laws instantiable elsewhere, but by the adaptation of locally situated practices to new local contexts' (Rouse, 1987, p. 125). Furthermore, public networks become 'the trading zones of knowledge' and 'politically normative commitment to public reasoning' is displaced by 'a "publicity machine" aimed at manipulating "public opinion" and transforming political actors into consumers' (Emden, 2013, p. 80). Under democracy, citizens 'learnt to make up their own minds and take responsibility for their own actions', they do not want 'patronising assumptions' and have their own vision on what they want to have in 'their society and their world' (Irwin and Michael, 2003, p. 1).

As scientific communities are linked to audiences and the public, scientific practice embedded in different publics, 'experimental knowledge circulates between different publics might also lead to an understanding of the way in which the boundaries between the public and the private are constantly redrawn' (Emden, 2013, p. 78). Citizens thus transform into suppliers of knowledge and policymakers into recipients.

Literature

Emden, C.J. (2013) Epistemic publics: On the trading zones of knowledge, in C.J. Emden, D. Midgeley (eds.) *Beyond Habermas: Democracy, knowledge and the public sphere*. New York: Berghahn, pp. 63–86.

Emden, C.J., Midgley, D. (2013) Beyond Habermas? From the Bourgeois public sphere to global publics, in C.J. Emden, D. Midgeley (eds.) *Beyond Habermas: Democracy, knowledge and the public sphere*. New York: Berghahn, pp. 1–16.

Entman, R.M., Bennett, W.L. (2001) Communication in the future of democracy: A conclusion, in W.L. Benett, R.M. Entman (eds.) *Mediated politics: Communication in the future of democracy*. Cambridge: Cambridge University Press, pp. 468–480.

Habermas, J. (2004) *Between facts and norms: Contributions to a discourse theory of law and democracy*. Frankfurt am Mein: MIT.

Habermas, J. (2005) *The structural transformation of the public sphere*. Cambridge: Polity Press.

Irwin, A., Michael, M. (2003) *Science, social theory and public knowledge*. Maidenhead: Open University Press.

Rouse, J. (1987) *Knowledge and power: Toward a political philosophy of science*. New York: Cornell University Press.

8 Different kind of science

Different actors producing science on GM crops

The introductory chapter has identified the increased engagement of civil society in the social management of science with the decreasing role of the state in the background. There is also another actor – business. Civil society and business each has produced its own science discourses which are worthy of further study.

The business sector has been brought into scientific research due to the capital-intense nature of modern science. That, in turn, has resulted in changes in the nature of science: academic science began to be replaced by industrial science (in the 1960s) and post-industrial science oriented to problem-solving in local contexts (Ziman, 1996). The difference between industrial and post-industrial sciences is in the substitution of 'market competition' by 'command management' of a small group of MNCs (Ziman, 1996, p. 76). Both led to the establishment of post-academic science which adopted a postmodern philosophy, as feared by some scientists such as Müller-Hill (1998). The spread of transnational corporate science, which is a 'business-science hybrid', has also raised issues of public interest and regulatory regimes for the intellectual ownership of research property (Glover, 2002). Corporate R&D laboratories conduct industrial research and produce new knowledge, so it is far too simplistic to see them 'as university research laboratories in exile as instruments of big business that manipulate once-pure scientists for corporate ends, or as second-rate research institutions' (Hounshell, 1988, p. xvii). Corporate management of research providing the basis for continuity of developing and maintaining laboratories was able to balance short-term needs with long-term strategies in research, which then led to decentralised research (Hounshell, 1988).

Some products developed by corporate scientists served well. Even DDT (dichlorodiphenyltrichloroethane), the Geigy product, was developed to fight insects, was useful in the World War II to address sanitary and hygiene limitations and proved to be an effective insecticide in agriculture until it was found to be toxic to the environment (Mellanby, 1992). The DDT case shows that even proven results might be contested in the future.

Corporations also provide funding to universities, and this gives ground for activists to accuse academic scientists of biased research. Natural scientists may find discomfort in this new expectation of not only discovering facts but also providing socially acceptable interpretations of these facts, lobbying for their further use and being aware of their public image. However, if scientists are not ready to operate in this new mode, they can be threatened by the loss of their ability to conduct research altogether. For example, Sir David Baulcombe, Head of the Department of Plant Sciences at the University of Cambridge, had to speak in front of farmers in Norfolk in the UK. While he was prepared to explain the scientific aspects of his research, he was confronted by the angry audience emotionally stimulated by an anti-GM activist (Baulcombe, 2014). It was a traumatic experience, but does that mean that academic scientists should avoid corporate funding? It is hardly possible. While universities continue to receive corporate funding and there are issues of corporate ownership rights over technologies, academic researchers are also involved in open technologies. For example, in the OpenPlant project, while there is still a need to protect the intellectual property of applications with potential commercial possibilities, its creators allow 'a family of generic lower-level tools that are largely free of IP constraints' to be freely shared to promote innovation in plant synthetic biology (OpenPlant, 2016).

Simultaneously, civil society has also become more involved in doing and discussing science. For example, an international NGO, the World Wildlife Fund (WWF), has since 1962 run its own research programmes aimed at the conservation of species threatened with extinction (WWF, 2015). It often commissions research by individual scientists and academic institutions (Schwarz, 2010). Within the WWF, specific scientific activities are accompanied with political lobbying and the education of general public, and the organisation enjoys its own pool of financial support based on mixed sources of funding. Increased interactive science communication with the public has been documented in another new concept – citizen science (CS) (Irwin, 1995). In certain cases, some academic research projects have benefited from including extra human and financial resources provided by concerned citizens. For example, the monitoring of changes in the migration of monarch butterflies in the US was framed into a network of projects run by volunteers with financial support from different foundations (Howard and Davis, 2015). While for some authors (Cohn, 2008) it has seemed natural to benefit from the inclusion of volunteers into scientific research, others have been more critical about its implications for academic science. For example, Riesch and Potter (2014) argued that 'overly grand aspirations for CS' can be misleading. Individual projects can illustrate the benefits of spreading knowledge, but on a systematic level, there will be a number of concerns of an ethical nature, such as research authorship and even fewer paid opportunities for professional scientists.

The promotion of CS brought a closer interaction of professional scientists with lay, non-scientific members of society to discuss science and make decisions over its applications. This has been reflected in the language and style in which these discussions are held. As has already been shown with the example of Prof. Baulcombe, in the debates between scientists and lay people about biotechnology and its GM products, rationality sometimes became opposed to emotionality. Here is another example:

> I went to a panel at the nearest high school with a green member of the state parliament. There were 500 people in attendance and it was packed. I was winning the argument, and suddenly [his opponent] started to scream and cry. So I said to her, "Don't you think we should stop being so emotional and be more objective/factual about this?" At that point a 50 year old lady in the audience stood up and said, "Mr [name], are you only a brain or do you actually have a heart in this issue, too?" That's when it became very clear to me that . . . the problem for the big corporations is that they are already anonymous and faceless, perfect target for activists, you can't win with the rational staff, have to show a human face.
>
> (Rao, 2009, p. 160)

Some members of the biotechnology community have recognised the current success of the anti-GM lobby in the debate (Schurman and Munro, 2010). Thus, one may argue that lay people seem to win the debate, at least in the case of GM, by better knowledge of public relations and playing on instinct and emotion, particularly fear, in public. The apt choice of rhetoric appropriate for a very specific local context has been crucial.

Framing the GMO debates in the Frankenstein rhetoric proved to be a decision which worked for the anti-GM lobby. Today the term 'Frankenstein science', named from the famous short story by Mary Shelley, refers to crazy, ill-fated scientific experiments leading to destruction, inaccurate science or pseudoscience.

The linking of GMOs with Frankenstein's creature was really just a matter of time. In 1977, Arthur Lubow in the *New Times* newspaper article raised concerns that 'modern Dr Frankensteins have found a way to create brand-new forms of life' (Lubow, 1977). Once the debate on GMOs focused on food, being set in the context of the bovine spongiform encephalopathy (BSE), commonly known as the mad cow disease crisis of the 1990s, a new term – 'Frankenfood' – appeared. It was coined by Paul Lewis, Professor of English at Boston College, in his letter to the *New York Times* in 1992 (Casetta and Tambolo, 2013).

The debate was then joined by Jeremy Rifkin, a founder of the Foundation on Economic Trends (FOET). An American activist and economist by training, Rifkin also consulted with European leaders at the highest level (FOET, 2013a). He led one of the first anti-GM campaigns, called the Pure Food Campaign, which opposed the commercial arrival of GM food, specifically the GM tomato Flavr Savr by Calgene, which was the first ever GM food product to reach consumers. Rifkin's initiative brought together 'a coalition of organic farmers and restaurateurs, consumer and environmental groups, and animal welfare organizations opposed to the use of genetic engineering in food' (FOET, 2013b). The campaigners distributed flyers featuring a dinosaur pushing a grocery basket labelled 'Biotech Frankenfoods'. Their message was that 'corporate science' could alter and create life forms with 'enormous and frightening' possibilities (Hamilton, 1993).

In his books, Rifkin has recognised the strong power and 'terrible nature' of the 'new science' represented by biotechnology to bring enormous changes to society and has questioned such consequences including such 'monstrous and unwarranted intrusion' into human lives (Rifkin, 1999, p. xii). He spoke about the potential risks for human health and the environment from such new technologies and elaborated the anti-corporate argument, especially its ownership aspect, by portraying business as gene-hunters trying to monopolise indigenous plant knowledge who are confronted by 'good guys' from NGOs. But he also touched upon the scientists whose faculties have received corporate funding (Rifkin, 1999, p. xii).

The Frankenstein rhetoric served well to present the role of science in developing new technologies. The key concepts were the ambivalence of modern technology and its uncertainty (Turney, 1998). Some have argued that with such a metaphor, the comparison of biotechnology with the Frankenstein story has done harm to the public image of technology, as it was not 'helpful in understanding and led to further astray' (Casetta and Tambolo, 2013). The fate of the genetically engineered tomato on both sides of the Atlantic Ocean was similar, since it was withdrawn from production and became unavailable to consumers.

According to Belinda Martineau, one of the Calgene biologists, during the hearing over Flavr Savr, there was 'not much meaningful listening' and almost no 'meaningful exchange' between the two sides (Martineau, 2001). As she reflected upon the beginning of the debates, she recognised the poor 'execution (or lack thereof) of bridging that information gap' between the opponents 'especially . . . on the part of the pro biotech camp'. In their explanations, they oversimplified, conveyed general, not-so-convincing ideas, such as there being 'no evidence that any of these products is unsafe' and did not refer to scientific publications (Martineau, 2001, p. XI). No wonder they could not win over public opinion.

The other successful tactic of the anti-GM movement is mimicry of scientific activities. This includes the involvement of people with a scientific background and production of reports that contain features of scientific reporting though tailored for non-specialist readers, written by activists to deliver a specific political message.

A good example of this are publications produced by the Earth Open Source, a British anti-GM NGO (Antoniou et al., 2012, 2014), authored by Dr Michael Antoniou, a biologist employed by professional science institutions, while Dr John Fagan holds a PhD in biochemistry and molecular biology from Cornell University, US. In 1994, the latter took a stand against genetic engineering, renounced his research grants and decided to dedicate his life to anti-GM activism (Fagan, 2007). Alongside them is social scientist Claire Robinson, who has a Master's degree, although her field is specified neither in the reports nor on the website. This fact has led some inquisitive readers to question their credentials (Griekspoor, 2014). However, in 2015, Claire Robinson's name disappeared from the Earth Open Source website.

At the same time, the pro-GM side equally involves people who are not all natural scientists by training. Mark Lynas, a former Greenpeace volunteer who became a supporter of GM crops, has transformed himself into an almost reputable scientist. He 'had not read not a single scientific paper on the subject' until his public coming out on GM crops (Forbes, 2011, para. 9). Since then, he has published a number of books, including *The God Species*, which has received positive reviews particularly for his factual knowledge and emotional presentation, and has been consulted by governments on his writings (Forbes, 2011). He is not seen as a scientist by either scientists or activists.

The boundaries between science and lay people have been blurred. In papers written by activists, the academic format – which can be considered heavy reading – is replaced with a style suitable for mass readership. The message communicated though is definite and clear and requires certain political actions from its readers (i.e. to ignore and protest against GM food or protest against the pro-GM lobby).

The science policy work of activists also mimics public democratic action such as public consultation. This raises questions of representation and public interest. An example of an alternative participatory exercise is the People's Report on GM, based upon separate juries that met in Hertfordshire and Tyneside in the summer of 2003 and deliberated the issue of GM crops. They were organised by a team from the Policy, Ethics and Life Sciences Research Centre (PEALS), University of Newcastle, who had been inspired by the UK government GM debate held in 2002. The report was presented as an alternative to the official debate and condemns 'the way in which the elected Government has merely paid "lip service" to public debate on such a major issue as GM' (PEALS, 2003, p. 2). The jury was funded by, among others, the Consumers' Association, Greenpeace, the Co-Operative Group and Unilever. Given the long-standing position of Greenpeace on the promotion of organic agriculture and opposition to GM crops, it is not surprising that the conclusions of the report list:

- A critique of current conventional agricultural practices based on high inputs of fertilisers and pesticides.
- A proposal for support systems for agricultural techniques that do not rely on artificial chemicals, such as organic farming.
- A call for incentives to encourage retailers to act in the interests of smaller and organic UK farmers, rather than to import food from abroad.
- A call for bodies that regulate new agricultural and food technologies to be made fully accountable to citizens, together with specific proposals for reform.

(PEALS, 2003, p. 2)

This leads us towards a general discussion regarding the nature of these civil society groups. Much has been said about the bottom-up approach and civil society activism such as self-help groups (Edwards and Fowler, 2002). However, it is also true that there are organisations of a supposedly non-profit nature that are run by professionals making money out of conducting certain lobbying activities. They may claim to act in the interests of humanity, but what is identified as being the common good and how it should be lobbied for is decided by a small group of people, not even a large pool of the organisation's volunteers. Greenpeace is an example of this situation (Zelko, 2013).

As shown earlier, in the postmodern context, science has been fragmented into different sciences: academic, corporate, citizen and activist science. As it becomes difficult in many cases to provide definitive answers, particularly to the long-term consequences resulting from the application of research which demands time, space and resources for the further study, they are used against each other in an attempt to win over the opposition in what are actually political campaigns.

Case studies of advocacy science

Case study 1: the Pusztai case

Dr Árpád Pusztai, a Hungarian biologist, spent most of his academic career working for the Rowett Research Institute in Aberdeen; his main area of expertise was plant lectins, a toxic substance naturally produced by plants to repel insects (Pusztai, 1991).

In August 1998, he appeared on a Granada Television programme called *World in Action*. In that public appearance, he felt it necessary to raise and share his concerns over GM foods in the context of a study conducted at the Rowett Institute aiming to transfer a snowdrop plant gene to a potato. From his much later interview in *The Guardian* newspaper, it appears that he was not prepared to deal with the complexity of mass media and possible consequences of making a strong claim about GM plants: 'I am an academic scientist. I've never been exposed to this . . . I'm really not a very media person' he said (Randerson, 2008, para. 8). The director of the Rowett Institute did not foresee the coming storm either and even called his subordinate colleague to congratulate 'on the modest way in which he had presented the evidence on the programme' (para. 10). The original study to which Pusztai referred in his interview was a comparison of the effects on rats from eating GM potatoes with lectin transferred from the snowdrop and non-GM potatoes (Randerson, 2008). In the interview, he referred to the limitations of testing procedures and stated, 'I find it's very unfair to use our fellow citizens as guinea pigs' (GM-Free, 1999, p. 4).

Suddenly, the media and public authorities took notice of that study. The stories in the media added to the confusion by referring to potatoes modified with a lectin transferred from the jack-bean which is poisonous to mammals. Interestingly, it is not possible to trace the original source of this to the media (Randerson, 2008).

Pusztai's boss, Prof. Philip James, who worked with the British government during the BSE crisis and produced a blueprint document outlining recommendations on how to regulate food safety issues, had to intervene, speak up for the reputation of the Institute and address the media storm. The Institute released a press release and its head, Prof. James, criticised Pusztai: 'My change in attitude was dramatic because I discovered that Pusztai . . .had never conducted the studies which he had claimed' (Randerson, 2008, para. 14). The Institute published an audit report. In response, Pusztai strongly denied this accusation but nonetheless lost his job at the Institute.

The immediate discussion of the Pusztai study was then followed by what GM opponents called a 'campaign of disinformation'. 'Faced with this onslaught on its policies, Tony Blair's government sprang to the defence of its biotech buddies with a stepped-up campaign of disinformation' (GM-Free, 1999, p. 5). The anti-corporate rhetoric accusing the Rowett Institute of receiving a £140,000 grant from Monsanto was also brought in (GM-Free, 1999, p. 4). In scientific circles, the affair produced a 'titanic battle of experts' (Fedoroff, 2011). First, Pusztai sent the audit report and his TV interview script to a number of scientists who, in turn, issued a memorandum in his support, aiming to 'remove the stigma of alleged fraud' and restore their colleague's scientific reputation (Lee and Tyler, 1999). Some, such as Prof. Pierzynowski, stressed that he did not find the audit report objective and that the whole incident was 'per se dangerous, not only for Dr. Pusztai, but generally for free and objective science' (Fedoroff, 2011, para. 6).

In October 1999, *The Lancet* published an article by Pusztai and Ewen which presented to academic readers his research on lectin from a snowdrop plant (Galanthus nivalis), concluding that 'the possibility that a plant vector in common use in some GM plants can affect the mucosa of the gastrointestinal tract and exert powerful biological effect' (Ewen and Pusztai, 1999, p. 1354). *The Lancet* received severe criticism from the Biotechnology and Biological Sciences Research Council for publishing the article, although its editor, Richard Horton, supported the publication and allegedly received 'an aggressive call' from Peter Lachmann, a former Vice-President and Biological Secretary of the Royal Society and President of the Academy of Medical Sciences (Flynn and Gillard, 1999). The reviews varied from rather sympathetic justificatory messages that Pusztai's aim was simply to show the necessity of careful testing (Rhodes, 1999) to a critique by John Pickett of Rothamsted Research, Harpenden, UK, who publicly denounced the journal for ignoring his advice to reject the paper (Enserink, 1999).

Biologists, such as Nina Fedoroff, published their critique of the research, blaming it for serious methodological flaws (Fedoroff, 2011; Fedoroff and Brown, 2004). At this point, to respond to the criticism,

Pusztai moved into the environmental movement and published his open letter from the website of the anti-GM NGO, LobbyWatch. His response was based on an explanation of how his case was mistreated by the Royal Society and the Rowett Institute. He provided further scientific details of his study and included a harsh personal critique of Nina Fedoroff of whom he returned the accusation of 'superficiality in scientific matters' (Pusztai, 2006). It is important to note that before this incident, Pusztai was not an activist or an official member of any civil group. He was, in his own opinion, 'strictly science-based', without ideology (Randerson, 2008, para. 6).

In April 1999, the Royal Society set up a working group to review its statement on GM plants for food use. It consisted of five prominent scientists. The purpose of the review was to clarify the ongoing debate on the safety of GM food started by Pusztai and to assist the Royal Society in developing its stance on the debate. Members of the group requested six independent reviewers across different disciplines to provide their assessment of the study. They also contacted the author to request additional data which he had indicated existed but never provided on the grounds that these were internal documents (Royal Society, 1999, p. 2).

The group did not find 'convincing evidence of adverse effects from GM potatoes' based on the fact that the study had 'technical limitations of the experiments and the incorrect use of statistical tests' (Royal Society, 1999, p. 1). The report reiterated the previous Royal Society statement that 'any over-arching body analyse the current regulations, giving particular consideration to whether long-term animal feeding studies are necessary to provide greater information on allergenicity or toxicity' and recommended that any study on GM food safety should be peer reviewed before publication (Royal Society, 1999, p. 5).

After his '150 seconds of TV "fame"', as he put it himself, Dr Pusztai remained within the GM debate but kept a lower profile. In 2008, in a letter to anti-GM activists, he wrote:

> I asked for a credible GM testing protocol to be established that would be acceptable to the majority of scientists and to people in general. 10 years on we still have not got one. Instead, in Europe we have an unelected EFSA[1] GMO Panel with no clear responsibility to European consumers, which invariably underwrites the safety of whatever product the GM biotech industry is pushing onto us. All of us asked for independent, transparent and inclusive research into

the safety of GM plants, and particularly those used in food. There is not much sign of this either. There are still 'many opinions but very few data'.

(cited in GMWatch, 2009)

In 2008, he insisted he was 'not a campaigner' or a member of any lobbying group (Randerson, 2008, para. 6). However, he received, together with his wife, the Stuttgart Peace Prize awarded annually by the German NGO Die AnStifter ('the Instigators'), to people or projects involved 'in a special way for peace, justice and world solidarity'. The news of the award was announced by another NGO, GMWatch (GMWatch, 2009).

The Pusztai affair is significant not only because it catalysed the debate over the use of GM crops but also provoked another debate about the very way scientific experiments are conducted, interpreted and communicated. The debate, moreover, went beyond the borders of Britain, spreading to countries with a natural suspicion of GMOs where it was well received, as the German award to Dr Pusztai shows. It can be argued that the impacts of the Pusztai affair can be found over a long-term period at European level. The European authorities adopted the Novel Foods Regulation in 1997 which also covered GM food. This led to national governments adopting their own measures. Austria initiated the process by being the first to ban GMOs in 1997, followed by Luxembourg, France, Greece and Germany (Heijden, 2010).

Case study 2: the Séralini case

Another example of biotechnology 'advocacy science' is the Séralini case. Gilles-Éric Séralini, a professor of molecular biology at the University of Caen, France, chairs the scientific board of the organisation CRIIGEN (Committee for Independent Research and Information on Genetic Engineering) which is technically an NGO.

Séralini and his research came to public attention in 2012. The article 'Long term toxicity of a Roundup herbicide and a Roundup-tolerant genetically modified maize' was published in *Food and Chemical Toxicology* in November 2012 (Séralini et al., 2013). The article was retracted 'after a thorough and time-consuming analysis of the published article and the data it reports' (Elsevier, 2012).

In March 2013, Séralini and his co-authors published their responses to critics in the same journal, repeating their arguments and accusing their reviewers of being biased in the GM debate: for either working as plant biologists developing GM plant patents or being paid by Monsanto

(Séralini et al., 2013). The article was republished in 2014 by *Environmental Sciences Europe* (Casassus, 2014).

The Séralini study examined the possible health effects of a Roundup-tolerant NK603 GM maize on the population of rats. The results presented suggested high toxic effects of the GM food given to the rats, including marked kidney deficiencies and high death rates (Séralini et al., 2012). The publication provoked academic debate, resulting in 18 letters to the editor of the journal and the original article being retracted. While the first letters presented a critique of the Séralini study from the methodological points of view (Ollivier, 2013), the later correspondence was not about the study itself but spoke more about social impacts of the study including accusations of the scientific community having links to Monsanto (John, 2014). Also voiced were the concerns of some scientists over 'a manipulation of the scientific process to achieve activist gains' (Folta, 2014). Interestingly, both authors, Brian John and Kevin Folta, are both scientists and activists but on opposite sides of the GM debate.

This is the greatest difference between the two case studies: Pusztai's publication was mostly criticised from the methodological point of view and Séralini's was situated in the discussion of where science and scientific communication stands, and its moral aspects. Thus, it can possibly be argued that in a relatively short period of time, 1998–2012, a large change in the perception of science occurred: it was now seen in postmodernist style as a social construct, including by natural scientists themselves.

The Séralini team delayed publication in order to match the release with a video entitled 'Are we all guinea pigs?', using similar rhetoric to Pusztai, and organised a media conference after which photographs of mice tumours appeared everywhere, including on American television (Arjo et al., 2013). As a result, the research received enormous coverage. Interestingly, before receiving the paper by Séralini et al., journalists were obliged to sign a non-disclosure agreement barring them from contacting any independent expert before publication (Lipponen, 2012).

As in the Pusztai case, when the Royal Society set up a working group to reassess the contested study, in 2012, the European Commission asked EFSA to review the Séralini study. EFSA's internal taskforce, chaired by the Director of Regulated Products, found the research 'to be inadequately designed, analysed and reported', criticising its authors for providing 'a limited amount of relevant information which failed to address the majority of the outstanding questions raised in the Authority's first statement' (EFSA, 2012).

To promote their research, Séralini and his supporters created a website called GMO Seralini which is 'owned and maintained by a group of concerned citizens and scientists' (GMO Seralini, 2015). The managing editor is Claire Robinson, the previously mentioned colleague of Michael Antoniou, who also maintained contact with Árpád Pusztai. The editor of Sustainable Pulse's GMO News website which is linked to the website, is Henry Rowlands who is another British anti-GM activist. The website is supported by NGOs CRIIGEN, Earth Open Source and GMO Evidence (GMO Seralini, 2015).

As in the case of the Pusztai study, the outcomes of the Séralini study were significant and soon spread throughout France. Some even argued that, inspired by the Séralini presentation, activists destroyed a GM soybean consignment in France in 2012. Russia and Kazakhstan placed a ban on imports of the GM maize used in the Séralini study, while Kenya banned imports of all GM food and Peru enacted a 10-year moratorium on GM crops (Arjo et al., 2013, p. 256).

Unlike Pusztai, who insisted on being 'strictly science-based' and without ideology, Séralini had developed his ideology before his research came into the public eye. He published several books in French which explain his views on GMOs to the general public. They are arguably examples of advocacy science. These are *Genetiquement Incorrect* ('Genetically Incorrect') (2003), *Ces OGM qui Changeant le Monde* ('These GMOs that change the world') (2004) and *Pesticides-OMG-Aliments Nous pouvons nous depolluer* ('Pesticides-GMOs-Food. We can delete pollution') (2010). *Après nous le déluge?* ('After us, the Deluge?') is co-authored with a botanist, Jean-Marie Pelt who conducted his fieldwork in Western Africa and Southern Asia. The authors claimed to be of the few dispersed voices on all continents who fight against the massacre of living beings and who want to give 'our concrete scientific experience so that you can judge the situation'. The situation they referred to is that humans will soon be incapable of living on the planet in good health, citing fertility issues 'because of genetic changes' and the loss of their citizenship to manage 'their own simple lives of a human' (Séralini and Pelt, 2008, p. 9). They also mentioned climate change, loss of biodiversity, and the use of natural resources. In their opinion, to reverse the situation, humanity faces a radical transformation 'to return to [the] satisfactory sanitary state' of the planet (Séralini and Pelt, 2008, p. 10). GM crops are presented as a biodiversity threat capable of killing wildlife (Séralini and Pelt, 2008, p. 77).

They asked scientists, particularly biologists, to review their role, which would arguably lead to merging science with activism:

> For forty years biologists played a very small role. Forget the man at the top of assembled cells, ignoring the landscape in which he moves, ignoring the planet in which he has a niche with plants and

animals. Outside the research, the world is faded. He does not study anything but genes, micro- and nanoparticles. Some scientists continue to project their fantasies of simplicity, like one gene = one protein = one function.

(Séralini and Pelt, 2008, p. 13)

They referred to reductionism in biology. In addition to scientists – reductionists, they identified another type of contemporary scientist – 'the interventionists those who by their own will change the state of nature – do not appear to regard the state of nature as a result of their ethics, it often does not interest them' (Séralini and Pelt, 2008, p. 78).

Although they did not claim science to be 'good or bad', they did 'judge the tree of its fruits'. So they criticised post-industrial science for 'establish[ing] a new religion' which is based upon 'political and economic powers' and expressed disapproval of 'scientists who are not in agreement of the impact of their activities paralysed by political will or comforted by their inertia' (Séralini and Pelt, 2008, p. 78). They questioned the moral authority of political leadership that approve GMOs and compared it with the nuclear power sector (Séralini and Pelt, 2008, p. 15).

As Séralini had expressed his negative views on GMOs prior to his study, this raises concern about bias in his participation and conduct of GM research. His publications call upon the action of the general public and are examples of political activism. While it is possible to assume that the Pusztai case has been accidental, the Séralini case was probably not. It could be due to conscious choice reflecting the social values of the scientist, and engagement with NGOs was not a last option but a thought through strategy to promote his study.

Nielsen (2001) suggested not mixing scientific research skills and personal values. He is right, but it seems difficult in practice. Yet, in the postmodern context of uncertainty about facts, different epistemologies and disagreements about values, when scientists get involved in a public debate or publicly share their professional findings and personal values, this is picked up by lobbying groups and becomes a big storm. In the two cases analysed, both scientists have lost at least some of their scientific credentials but gained influence among activists.

The case of Anne Glover

A contrasting case is that of Prof. Dame Anne Glover whose position as Chief Scientific Adviser (CSA) to the president of the European Commission was terminated after lobbying by NGOs in Brussels. This happened most probably because of her support for GM crops – she offered the opinion that according to the scientific consensus, GM crops are safe (Briggs, 2015). Unlike Pusztai and Séralini, Glover lost her position to influence public policy as a result of political lobbying, although she retained her academic credentials.

Glover is a Scottish scientist who studied biochemistry at the University of Edinburgh. In 1982 she obtained her PhD from the University of Cambridge after which she moved to the University of Aberdeen to carry on her microscopic studies on soil. She created bioluminescent bacteria, a powerful technique to monitor soil health and pollution levels. She also set up a company called Remedios to commercialise her invention (Martynoga, 2015). Yet her most famous position was as Chief Scientific Adviser to the president of the European Commission from 2012–2014. When she reflected on her time as CSA, it was quite an outstanding experience which she compared to a rewarding journey with 'elements of Quixote, Kafka and Macondo' (Glover, 2015, p. 60). There was a clash of the idealistic Habermasian-style vision of scientific knowledge held by natural scientists and the complex reality of multi-actors within the European Union who would rather insist on their own interpretation of scientific knowledge and lobby their specific interests. In this context, the reference to Don Quixote fighting windmills is appropriate.

The position of CSA was established by José Manuel Barroso, president of the European Commission, in late 2011. Glover was nominated as the first CSA. According to Glover, the post 'gained high public visibility and quickly developed into a champion for the use of scientific evidence by the Commission, both as a "voice of science" at the political level and as a "scientific ombudsman' for third parties" (Glover and Muller, 2015, p. 37). The CSA was responsible for 'establishing for the first time a network of government science advisers across the Member States (the European Science Advisers Forum) and supporting the new EU Agencies Network of Science Advisers, which aimed at better coordination' (Glover and Muller, 2015, p. 37).

She faced the issue of limited staff support (a secretary and part-time available expert assisted her) and a small budget, with a further challenge of being 'cut off from vital information' as she was not included in important meetings of the Directors-General (Glover, 2015, p. 63). As a result, internal information would reach Prof. Glover accidentally, and there were definite gaps in communication between the CSA and the Commission (Glover, 2015, p. 63).

That was contradictory to her vision – she had hoped to leave a long-lasting impact on improving science communication in Europe. She advocated for both the unbiased presentation of scientific information and transparency in communication:

> As a matter of principle, evidence needs to be procured from the widest possible range of sources and used and communicated in an unbiased manner. There must be complete transparency on what questions are used to request evidence, where the evidence comes from, how it is analysed, how it is prioritised and what the conclusions are.
>
> (Glover and Muller, 2015, p. 37)

Glover insisted that 'the evidence-gathering process itself must not be compromised by a politically desired outcome' (Glover and Muller, 2015, p. 37). She saw the function of CSA as serving as a 'scientific ombudsman', ensuring that the president or other commissioners can quickly access scientific advice when required, and ensure that the voice of science is 'in the room' when political decisions are taken (Glover and Muller, 2015, p. 39).

She was against a technocratic approach to evidence and in support of transparency to citizens who should know 'when economic, social, ethical or political imperatives override the scientific evidence' and 'avoid the impression that science and technology "dictate" a particular outcome' (Glover and Muller, 2015, pp. 33, 40).

The clash of visions became visible when the CSA and president of the European Commission received an open letter signed by a group of scientists arguing that the Commission ignored scientific knowledge about endocrine disruptors. The opposing side also submitted their letters. Prof. Glover chose not to support any side and 'found this public discussion to be rather unhelpful for science' and replied to three experts in the field to discuss the areas in question. This 'triggered the fury of NGOs and journalists' hoping to present the case as lobby interests and accused the CSA for the delay in the decision on European legislation on the endocrine disruptor (Glover and Muller, 2015, pp. 71–72).

This was 'not the only hornet's nest' she dared to stir. As she spoke about the science on GMOs, Corinne Lepage, the French Member of the European Parliament (MEP), together with the previously mentioned Gilles-Éric Séralini, held a public conference in November 2013. The MEP called on Prof. Glover to resign from her post, accused her of 'convey[ing] a one-sided view on genetic engineering' and being paid by Monsanto (Glover and Muller, 2015, p. 73). Later on, Prof. Glover proved that this was a false statement and Ms Lepage was forced to retract it (Glover and Muller, 2015, p. 73).

In 2014, Pesticide Action Network and Corporate Europe Observatory, both NGOs, sent a request to access all correspondence held by the CSA office on the topic of endocrine disruptors and pesticides, GMOs and the precautionary principle. The requests were to be addressed within 15 working days and pressure was placed on the CSA. In May 2014, Glover invited several NGO representatives to a meeting to discuss how the CSA role could be improved. Later that year, in July, nine NGOs asked the future president of the European Commission, Jean-Claude Juncker, to abolish the role of the CSA on the grounds that 'the post of Chief Scientific Adviser is fundamentally problematic as it concentrates too much influence in one person, and undermines in-depth scientific research and assessments carried out by or for the Commission directorates in the course of policy elaboration' (Corporate Europe Observatory, 2014, para. 2). One of the signatories was the already mentioned Claire Robinson, representing NGOs GMWatch and Earth Open Source (Corporate Europe Observatory, 2014).

To her disappointment, the post was abolished and she never met with the new president (Glover, 2015, p. 76). She assumed that the opposition from NGOs to the role of CSA 'at least in part, seemed motivated by [her] public remarks regarding the evidence around GM technology' (Glover, 2015, p. 78).

To compare Glover's view, it is interesting to read the almost apologetic comments from the Greenpeace representative explaining that only one think tank submitted an open letter asking for the abolishment of the CSA post based on the argument 'specifically because they disagree with Professor Glover's advice on genetically modified crops and organisms'. According to Greenpeace, the point of the letter submitted by nine NGOs 'was about the CSA's presentation to the media of a supposed scientific consensus, in contrast to the secrecy surrounding the actual science advice she was giving' (Parr, 2015, p. 86). The joint letter by NGOs stated:

> To the media, the current CSA presented one-sided, partial opinions in the debate on the use of genetically modified organisms in agriculture, repeatedly claiming that there was a scientific consensus about their safety whereas this claim is contradicted by an international statement of scientists (currently 297 signatories) saying

that it 'misrepresents the currently available scientific evidence and the broad diversity of opinion among scientists on the issue'.

(Corporate Europe Observatory, 2014, para. 3)

In the end, Prof. Glover returned to Scotland to continue her successful academic career and the European Commission appointed a team of scientists instead of a single CSA. In 2015, she wrote that there is no 'problem with the evidence per se', but it is a problem with how evidence is used and communicated (Glover and Muller, 2015, p. 33).

Note

1   The European Food Safety Authority

Literature

Antoniou, M., Fagan, J., Robinson, C. (2012) GMO *myths and truths: An evidence-based examination of the claims made for the safety and efficacy of genetically modified crops.* 1st Edition. London: Earth Open Source.

Antoniou, M., Fagan, J., Robinson, C. (2014) GMO *myths and truths.* 2nd Edition. London: Earth Open Source.

Arjo, G., Portero, M., Pinol, C., Vinas, J., Matias-Guiu, X., Capell, T., Bartholomaeus, A., Parrott, W., Christou, P. (2013) Plurality of opinion, scientific discourse and pseudoscience: An in depth analysis of the Seralini et al. study claiming that Roundup TM ready corn or the herbicide Roundup TM cause cancer in rats. *Transgenic Research*, 22: 255–267.

Baulcombe, D. (2014) Interview. Cambridge, 6 February 2014.

Briggs, H. (2015) Ex-EU science chief adviser Anne Glover: GM tech 'is safe'. *BBC News*, 2 February 2015. www.bbc.co.uk/news/science-environment-31091821

Casassus, B. (2014) Paper claiming GM link with tumours republished. *Nature News*, 24 June 2014. www.nature.com/news/paper-claiming-gm-link-with-tumours-republished-1. 15463

Casetta, E., Tambolo, L. (2013) That frightening Frankenmetaphor!, in N. Michaud (ed.) *Frankenstein and philosophy.* Chicago: The Open Court Philosophy, pp. 49–58.

Cohn, J. P. (2008) Citizen science: Can volunteers do real research? *BioScience*, 58 (3): 192–197.

Corporate Europe Observatory (2014) The position of the Chief Scientific Advisor to the President of the European Commission. A letter to Jean-Claude Juncker, President-elect of the European Commission. 22 July 2014. https://corporateeurope.org/power-lobbies/2014/07/position-chief-scientific-advisor-president-european-commission as viewed 18.10.2016.

Edwards, M., Fowler, A. (2002) *The Earthscan reader on NGO management.* London: Earthscan.

EFSA (2012) Statement of EFSA. Final review of the Séralini et al. (2012a) publication on a 2-year rodent feeding study with glyphosate formulations and GM maize NK603 as published online on 19 September 2012 in Food and Chemical Toxicology. *EFSA Journal*, 10(11): 2986.

Elsevier (2012) RETRACTED: Long term toxicity of a Roundup herbicide and a Roundup-tolerant genetically modified maize. www.sciencedirect.com/science/article/pii/S0278691512005637 as viewed 20.08.2016.

Enserink, M. (1999) Transgenic food debate: *The Lancet* scolded over Pusztai paper. *Science*, 286(5440): 656.

Ewen, S.W., Pusztai, A. (1999) Effect of diets containing genetically modified potatoes expressing Galanthus nivalis lectin on rat small intestine. *Lancet*, 354: 1353–1354.

Fagan, J. (2007) A science-based, precautionary approach to the labelling of genetically engineered foods. www.psrast.org/jflabel.htp as viewed 23.06.2014.

Fedoroff, N. (2011) Pusztai's potatoes: Is 'genetic modification' the culprit? *AgBioWorld*. www.agbioworld.org/biotech-info/articles/biotech-art/pusztai-potatoes.html as viewed 20.08.2015.

Fedoroff, N., Brown, N.M. (2004) *Mendel in the kitchen: A scientist's view of genetically modified foods*. Washington, DC: Joseph Henry Press.

Flynn, L., Gillard, M.S. (1999) Pro-GM food scientist 'threatened editor'. *The Guardian*, 1 November 1999. www.theguardian.com/science/1999/nov/01/gm.food

FOET (2013a) Jeremy Rifkin. www.foet.org/JeremyRifkin.htm as viewed 20.08.2015.

FOET (2013b) Pure food campaign. www.foet.org/past/PureFoodCampaign.html as viewed 20.08.2015.

Folta, K. (2014) Letter to the editor. *Food and Chemical Toxicology*, 65: 392.

Forbes, P. (2011) The god species by Mark Lynas: Review. *The Guardian*, 20 July 2011. www.theguardian.com/books/2011/jul/20/mark-lynas-god-species-review

Glover, A. (2015) A moment of magic realism in the European Commission, in J. Wilsdon, R. Doubleday (eds.) *Future directions for scientific advice in Europe*. Cambridge: CSAP, pp. 60–81.

Glover, A., Muller, J.M. (2015) Evidence and policy in the European Commission: Towards a radical transformation, in J. Wilsdon, R. Doubleday (eds.) *Future directions for scientific advice in Europe*. Cambridge: CSAP, pp. 33–41.

Glover, D. (2002) Transnational corporate science and regulation of agricultural biotechnology, *Economic and Political Weekly*, 37 (27): 2734–2740.

GM-Free (1999) *Pusztai potatoes: The Chernobyl of biotech*. GM-Free, April 1999. Skelmersdale: KHI Publications, pp. 4–5.

GMO Seralini (2015) About. www.gmoseralini.org/about-us/ as viewed 20.08.2015.

GMWatch (2009) Pusztai to receive Stuttgart Peace Prize. www.gmwatch.org/news/archive/2009/11801-pusztai-to-receive-stuttgart-peace-prize as viewed 20.08.2015.

Griekspoor, P.J. (2014) Checking up on Earth Open Source's anti-GMO stance, farm progress. 26 May 2014. http://farmprogress.com/blogs-checking-earth-open-sources-anti-gmo-stance-8549 as viewed 20.08.2015.

Hamilton, J. (1993) Who's afraid of Jurassic Park? Biotech ought to be. *Bloomberg Business*, 6 June 1993. https://www.bloomberg.com/news/articles/1993-06-06/whos-afraid-of-jurassic-park-biotech-ought-to-be

Heijden, H.A. van der (2010) *Social movements, public spheres and the European politics of the environment: Green power Europe?* Basingstoke: Palgrave MacMillan.

Hounshell, D.A. (1988) *Science and corporate strategy Du Pont R&D 1902–1980*. Cambridge: Cambridge University Press.

Howard, E., Davis, A.K. (2015) Investigating long-term changes in the spring migration of Monarch butterflies (Lepidoptera: Nymphalidae) using 18 years of data from journey

North, a citizen science program. *Annals of the Entomological Society of America.*, 108(5): 664–669.

Irwin, A. (1995) *Citizen science.* London: Routledge.

John, B. (2014) Letter to the editor. *Food and Chemical Toxicology*, 65: 391.

Lee K., Tyler, R. (1999) International scientists raise concerns over genetically modified food. British Labour government rushes to defend biotech industry. World Socialist Web Site. 17 February 1999. www.wsws.org/en/articles/1999/02/food-f17.html

Lipponen, S. (2012). EUSJA Statement on embargoes and manipulation. European Union of Science Journalists Association. www.eusja.org/eusja-statement-on-embargoes-and-manipulation/ as viewed 20.08.2015.

Lubow, A. (1977) Playing god with DNA. *New Times*, 7 January 1977, 48.

Martineau, B. (2001) *First fruit: The creation of Flavr Savr and the birth of biotech food.* New York: McGraw Hill.

Martynoga, B. (2015) Dame Anne Glover. *Financial Times*, 13 November. www.ft.com/content/55e0dcf6-830e-11e5-8e80-1574112844fd as viewed 17.10.2016.

Mellanby, K. (1992) *The DDT story.* Farnham: British Crop Protection Council.

Müller-Hill, B. (1998) The different faces of science: Is genetics a social construct?, in K. Bayertz, R. Porter (eds.) *From physico-theology to biotechnology: Essays in the social and cultural history of biosciences: A festschrift for Mikulas Teich.* Atlanta: Editions Rodopi, pp. 40–52.

Nielsen, L.A. (2001) Science and advocacy are different – and we need to keep them that way. *Human Dimensions of Wildlife*, 6: 39–47.

Ollivier, L. (2013) Comment on Séralini, G.-E., et al., long term toxicity of a Roundup herbicide and a Roundup-tolerant genetically modified maize. *Food and Chemical Toxicology*, 53: 458.

OpenPlant (2016) Data sharing. http://openplant.org/research/data-sharing/ as viewed 07.07.2016.

Parr, D. (2015) Why it made sense to scrap the post of the chief scientific adviser, in J. Wilsdon, R. Doubleday (eds.) *Future directions for scientific advice in Europe.* Cambridge: CSAP, pp. 82–89.

PEALS (2003) *The people's report on GM.* Newcastle upon Tyne: Bioscience Centre.

Pusztai, A. (1991) *Plant lectins.* Cambridge: Cambridge University Press.

Pusztai, A. (2006) Pusztai replies to Fedoroff. Lobbywatch.org, 14 March 2006. www.lobbywatch.org/archive2.asp?arcid=6338 as viewed 20.08.2015.

Randerson, J. (2008) Arpad Pusztai: Biological divide. *The Guardian*, 15 January 2008. www.theguardian.com/education/2008/jan/15/academicexperts.highereducationprofile

Rao, H. (2009) *Market rebels: How activists make or break radical innovations.* Woodstock: Princeton University Press.

Rhodes, J.M. (1999) Genetically modified foods and the Pusztai affair. *BMJ*, 318(7193): 1284.

Riesch, H., Potter, C. (2014) Citizen science as seen by scientists: Methodological, epistemological and ethical dimensions. *Public Understanding of Science*, 23(1): 107–120.

Rifkin, J. (1999) *The biotech century: Harnessing the gene and remaking the world.* New York: Penguin Putnam.

Royal Society (1999) Review of data on possible toxicity of GM potatoes. London, June 1999. https://royalsociety.org/~/media/Royal_Society_Content/policy/publications/1999/10092.pdf as viewed 20.08.2015.

Schurman, R., Munro, W.A. (2010) *Fighting for the future of food: Activists versus agribusiness in the struggle over biotechnology.* Minneapolis: University of Minnesota Press.

Schwarz, E. (2010) Interview with Eugeny Schwarz. Moscow, 27 March.

Séralini, G.E., Clair, E., Mesnage, R., Gress, S., Defarge, N., Malatesta, M., Hennequin, D., Spiroux de Vendomois, J. (2012) Long term toxicity of a Roundup herbicide and a Roundup-tolerant genetically modified maize. *Food and Chemical Toxicology*, 50: 4221–4231.

Séralini, G.E., Mesnage, R., Defarge, N., Gress, S., Hennequin, D., Clair, E., Malatesta, M., Spiroux de Vendomois, S. (2013) Answers to critics: Why there is a long term toxicity due to a Roundup-tolerant genetically modified maize and to a Roundup herbicide. *Food and Chemical Toxicology*, 53: 476–483.

Séralini, G.E., Pelt, J.M. (2008) *Après nous le déluge?* Paris: Flammarion.

Turney, J. (1998). *Frankenstein's footsteps: Science, genetics and popular culture*. New Haven: Yale University Press.

WWF (2015) *The evolution of WWF*. Gland: WWF.

Zelko, F. (2013) *Make it a green peace!: The rise of countercultural environmentalism*. Oxford: Oxford University Press.

Ziman, J. (1996) 'Postacademic science': Constructing knowledge with networks and norms. *Science Studies*, 1: 67–80.

9   Changing sides in the debate

In every debate, there are three positions: proposed (in favour), oppositional and neutral. In formal political debate, such as in parliamentary debate, when speakers put forward a proposal or 'question' it is also referred to as a motion. A motion can be debated, amended, superseded, adopted, negated or withdrawn; it can be also restricted to a set period of time and be set aside (Marleau and Montpetit, 2000).

In general, informal political debates, such as those on GM crops, there is no formal set of procedures, and it is interesting to reflect upon how the debate has developed in terms of people choosing which side and what arguments to support. Particularly, it is interesting to look at what it takes to change sides in the debate.

Political scientists have cautioned against assessing the public as competent and rational in its participation in political processes and 'making a reasoned choice' (Lupia and McCubbins, 2012, p. 49).

When people choose sides, ideally, they receive the relevant information and use it to make their decision. In the case of a complex modern issue, neither the general public nor policymakers screen out information and use it effectively, making a logical decision only under favourable conditions which include 'intrinsically simple' tasks, the availability of 'helpful capabilities or dispositions, and competent institutions (Kuklinsky and Quirk, 2012, p. 167). When citizens are asked to decide upon a complex issue with many uncertainties, they tend to use stereotypes to fill in missing information and more often relate to the testimony of others and not their own personal experience to access such information (Kuklinsky and Quirk, 2012, p. 167).

Social psychology studies explaining how to win public opinion in debate distinguishes three elements: attitude consistency, which is the predictability of the individual's position on the issue; stability of opinions on the issue; and its conceptualisation, that is, scoring reasons for choosing to support or object to the issue (Sniderman et al., 1991, p. 2). In everyday life, people can change their opinion and this is not a big deal. However, within political debate it might become 'the aim of political argumentation', and 'saying that you have changed your position on an issue need not be the same as changing it' (Sniderman et al., 1991, p. 236). Depending on how contested and debated the issue is and how much public visibility is attributed to the person changing their position, all that can increase attention on the act of changing sides in the debate and can potentially change the course of the debate itself.

In the previous chapters, we have discussed how members of NGOs choose their ideology in order to participate in public debate. It is evident that there are two sides – proponents and opponents to the use of GM crops, and on special occasions they switch sides, attracting media attention and possibly affecting the course of the debate. There is also one more, which we referred to as withdrawn. In this chapter, we will discuss what happens when participants of the GM debate decide to change their view on transgenic crops.

Actors participating in the debate and their sides

It is important to identify the actors involved in the GM debate. They include scientists, civil society representatives, members of NGOs or activists, policymakers and representatives of business, farmers and consumers who also either form their own NGOs or support activist campaigns and organisations. Together, these form a wide spectrum of participants. Each group has different interests and intentions in influencing policies on GM crops. The major debates are taking place between scientists, often among themselves, and activists, and frequently involve politicians. Gaudilliere (2006) separated two levels of debate: the scientific debate and the 'lay' debate which includes other scientists beyond molecular biology and other interested parties of 'lay people' (i.e. not scientists). The research by Cook et al. showed how scientists tend to distance themselves from both the public and opponents of GM who consist of the group that would include some sections of the media, as well as campaigning NGOs (Cook et al., 2004). Indeed, the interactions of science with other public actors have not been smooth. Scientists are also confronted with the call to present and explain their research to the wider public which has already been incited to negative opinions (Baulcombe, 2014). The feelings of 'wasting time' and withdrawal from the conversation that form the result of these debates by scientists are understandable (Baulcombe, 2014; Dobermann, 2015). However, more and more scientists choose to participate in the lay debate, as was shown in the previous chapter. In many cases, the same people who represent the scientific community become activists and vice versa.

The media acts as a voice through which public opinion can be influenced and it uses direct and indirect ways to engage with policy decision-makers. Farmers and consumers, as individual bodies, are often referred to by both opponents and proponents of GM crops as the constituencies that would benefit from the results of such lobbying.

Politicians have also participated in the debate. For example, the President of the European Commission, Jean-Claude Juncker, in his opening statement upon taking office, claimed that his plans to review the legislation on the authorisation of GMOs were aimed 'to give the majority view of democratically elected governments at least the same weight as scientific advice, notably when it comes to the safety of the food we eat and the environment in which we live' (Juncker, 2014, pp. 11–12). Former International Coordinator of Greenpeace International's Genetic Engineering Campaign, Benny Haerlin founded his own NGO called Gen-ethic Network in Germany in 1987. In 1979, he had joined the anti-nuclear movement and was a member of the Green Party in the European Parliament.

Table 9.1 Changing positions in the debate

Pro-GM	Withdrawn	Anti-GM
Stewart Brand	Oxfam	Thierry De Vrain
Jens Katzek	WWF	Government of Burkina Faso
James Lovelock		
Mark Lynas		
George Monbiott		
Bill Nye		
Paul Temple		

As a politician, he understands the value of 'effective networking' and continues to lobby in Brussels (Haerlin, 2014). Thus, it is quite possible to suggest that the frontier between activism and politics is rather porous.

There is also a geographic context to the debate. Europe is widely known for opposing the cultivation of GM crops in the region, as shown in Chapters 2 and 4. Many of these concerns are rooted in the past and relate to European farming interests. Thus, a large amount of participants in the debate who side against GMOs are from European NGOs or are connected to Europe. However, possibly the most outstanding anti-GM activist is Indian, Dr Vandana Shiva.

Table 9.1 lists some of the activists and organisations whose position on GM crops has changed. As can be seen, there are more cases of activists changing from the anti-GM side to the pro-GM side. Dr Vrain, on the other hand, is an even rarer example since he changed his position from working as a research scientist for Agriculture Canada, 'address[ing] public groups and reassur[ing] them that genetically engineered crops and foods were safe', to educating the public about concerns about the danger of GM crops to environment and health (Vrain, 2013).

Eco-heretics and Golden Rice

The term 'eco-heretics' is a new one, describing a small group of environmentalists who have changed their opinions on a number of matters, such as the use of nuclear energy and GM crops. Most have done so in regards to GM crops while discussing the Golden Rice project.

Professors Peter Bayer and Ingo Potrykus started to work on the Golden Rice project in 1992 with the hope of creating a new variety of rice reinforced with Vitamin A to fight the condition of Vitamin A deficiency (VAD) which is widespread among women and children in developing countries. The first trial was conducted in 2004. The following year, with the contribution of the Syngenta Foundation, a new version of the seeds was developed. In 2008, the Bill and Melinda Gates Foundation supported the project, with clinical trials being completed in 2009 (P. Moore, 2014). The research met two challenges along the way: the intellectual property rights ownership of biotech corporations and opposition from the environmental movement.

The first challenge was met by free licences donated by the involved companies. Syngenta also set up a foundation, 'Humanitarian Golden Rice Project', to start a dialogue within the public sector about the research and distribution of the crop. It was announced that public rice institutions would develop locally adapted Golden Rice varieties. Farmers with smallholdings were to receive new seeds free of charge, while big farmers were to pay a licence fee (Sahai, 2004, p. 4612). Thus, at least three of the regular criticisms of GM crops were not applicable to Golden Rice: the involved corporations were not making a profit, poor farmers had access to the seeds at no cost and local varieties were being promoted. However, Syngenta's further conditions over the use of Golden Rice drew criticism since the company left only one variety of genetically transformed rice line and appointed Gerard Barry, a former Monsanto employee, as the Golden Rice coordinator (Sahai, 2004, p. 4613; King et al., 2011).

To overcome the opposition of environmental NGOs to GMO crops, Potrykus approached his natural enemy – Greenpeace. According to a letter released to the public, he initially received what seemed to be a positive response from Greenpeace's campaigner, Benedikt Haerlin (Potrykus, 2001). However, according to interview, Haerlin has never seriously considered accepting Golden Rice and suggested to Potrykus an alternative – 'eating more carrots'. He mentioned that despite his official step out from Greenpeace immediately after the incident with Potrykus, he continued to collaborate with the organisation for a few more years (Haerlin, 2014). In his letter, Potrykus warned Greenpeace from planning to destroy his field trials, accusing it of 'contributing to a crime against humanity' (Potrykus, 2001). In August 2013, activists supported by Greenpeace destroyed the Golden Rice trials in the Philippines (Lomborg, 2013).

A former Greenpeace member, Patrick Moore, then stepped into the campaign for Golden Rice. Moore was born in 1947 in Canada and is one of the co-founders of Greenpeace. He joined his fellow first members of Greenpeace in 1971 and took part in its first campaigns, such as a boat rally in the Pacific. He claimed to be the only one among the group to hold a university degree in fundamental sciences; the others were social scientists (M. Moore, 2014). Relatively soon, he started to have disagreements about the methods and campaigns led by the organisation. In both his talks and his book, he explains that he wanted a more 'sensible, science-based approach to environmentalism'. He moved into the consulting business that he had previously attacked and worked on social corporate responsibility issues (M. Moore, 2014).

Moore's main argument is based on earlier claims by Potrykus that Greenpeace has conducted a 'crime against humanity' by rejecting Golden Rice. Moore has referred to the Rome Statute of the International Criminal Court (ICC) (P. Moore, 2014) which deals with 'the most serious crimes of concern to the international community as a whole' (Preamble, ICC, 2002). It describes the procedures of prosecuting criminals and protecting their victims, including the established principles of the international law of armed conflict (ICC, 2002, Art. 21). It is his contention that the actions of Greenpeace are intentional and its campaign against Golden Rice is a 'systematic attack directed against any civilian population'. In his view, following the rules of the Statute, Greenpeace should be taken before the court and the protection of the humanitarian rights of the victims, in this case individuals suffering from VAD, meaning full access to Golden Rice (P. Moore, 2014).

Upon learning about the controversy of the Golden Rice case, Moore registered in September 2013 an NGO called 'The Allow Golden Rice Society' in his native Canada together with his brother Michael. The aim of the organisation is

> to end the active blocking of Golden Rice by Greenpeace and other organizations who claim that it is either of no value or that it is a detriment to human health and the environment . . . through direct public action, media communications and coalition-building.
>
> (P. Moore, 2014)

In 2014, the Society, effectively meaning the Moore brothers, completed three public campaigns (one in Canada, two in Europe), and in 2015, they toured the Philippines, Bangladesh and India. The campaigns included public lectures, media briefings, interviews and public manifestations outside Greenpeace offices. The London march against Greenpeace in early 2014 is possibly the first example of protests held against the organisation (AllowGoldenRiceNow, 2016).

Moore has been called by his former colleague, Bob Hunter from Greenpeace, an 'eco-Judas' but the two have remained on good terms (Zelko, 2013, p. 355). Technically, Moore has not changed his mind against GM crops, because he was not in Greenpeace when the GM crops issues came about. But he has changed his opinion about the organisation that leads the anti-GM rhetoric and as such gives a precedent of radical environmentalists switching sides.

Such cases are rare, although Moore is not the only example of an influential environmentalist who has gone against the mainstream of the green movement. The 'environmental heretics' include environmental journalist George Monbiot, whose initial remarks about GM as being 'the biggest threat to future supplies' were cited by GM-Free magazine (GM-Free, 1999); Mark Lynas, another former Greenpeace activist who used to destroy GM trials; scientist and science writer Stewart Brand; and James Lovelock, the founder of the Gaia theory (Connor, 2014). Harrison-Dunn added to the list Jens Katzek, a former anti-GMO campaigner for Friends of the Earth, who joined the Golden Rice Humanitarian Board (Harrison-Dunn, 2014). Many attracted media attention for their anti-mainstream claims: James Lovelock, whose ideas were discussed in Chapter 7, wrote a public letter in which he 'bow[ed] [his] head in shame at the thought that our original good intentions should have been so misunderstood and misapplied', referring to the Green movement rejecting all energy sources, except renewable ones (Lovelock, 2012).

While Golden Rice's initial design was made for a humanitarian cause, it is not the only reason that caused 'the heretics' to change their minds about GM technology in agriculture. A broader factor seems to be climate change and the search for mitigation of global warming. They have considered GM crops, together with nuclear energy, to be less harmful than fossil fuels which contribute greatly to emissions, and conventional agriculture with the heavy use of pesticides allowing for the economising of resources, such as water and soil, while still providing for the majority of the world population (Lovelock, 2009; Lynas, 2011a). While discussing food security from the point of what seems to be economic rationality, they offered a serious change in the environmental paradigm. Moore and Lovelock suggested that people's needs – in this instance, the need for food – were to be taken inseparably from those of the Earth (Lovelock, 2012). Unlike the traditional environmental view which sees humans as a threat to nature, they believe humans to be part of nature and thus should have our needs satisfied, particularly if technology provides the means to do so in a less environmentally harmful way (P. Moore, 2014).

Of course, the mainstream environmental movement does not easily accept such new arguments. The main GM opponents disapproved of Golden Rice. Greenpeace members and Vandana Shiva have criticised Golden Rice, using the same critique as for other GM crops, while adding the discussion of the possible nutritional failures of Golden Rice (King et al., 2011; Greenpeace, 2013; Shiva, 2001). There were also attempts to disqualify the credibility of these 'green heretics', narrowing the debate towards neo-liberal venality. Since Moore has worked for business, including timber companies, he is an outcast from environmentalists, particularly those with hardcore anti-GM views. In addition, allegations were made that Lynas worked for EuropaBio, the European biotech business association, but he was proved to have no connection to it (Lynas, 2011b).

Monbiot, who agreed to eat GM food (2002), in his critique of the BBC film entitled 'What the Green Movement Got Wrong' about Brand and Lynas, the 'green heretics', warned about the danger of demonising environmentalism altogether (Monbiot, 2010).

The coming out of Mark Lynas

Mark Lynas is another eco-heretic and another former Greenpeace member who changed his views on GM crops. Lynas was born in the UK in 1973 and holds a degree in Modern Politics and History from Edinburgh (Lynas, 2014a). Seventeen years ago, Lynas was with Greenpeace and even went to Bangladesh where he met Vandana Shiva who he admired and referred to as 'a guru type'. According to him, at that time Greenpeace worked against Bt eggplant, or brinjal in Hindi (Lynas, 2014a).

Lynas recognises that he was involved in GM crop vandalism and while he was not a co-founder, he was among the first members of Greenpeace to attack Monsanto. He himself organised campaigns, hired buses and attacked farmers' crops. He also attended the Genetic Network, an informal group (Lynas, 2014a). Lynas has gone on to write two bestselling books on climate change and international development: *Six Degrees: Our Future on a Hotter Planet* (2008) and *The God Species: Saving the Planet in the Age of Humans* (2011a).

To the question of whether he was inspired by Patrick Moore, he answered that he was not and that it had been his own decision to leave Greenpeace because of his changing views on GM crops. In regard to Moore, Lynas had concerns about his involvement 'as a frontman' because of his connections with business from which he 'made money'. At the same time, Lynas was not satisfied with how Greenpeace promoted its ideology to members; it seemed to him that the organisation 'ha[d] a monopoly of virtue'. He also gives a clue that some other members of Greenpeace are quite close to his own viewpoint, but keep silent because they need to keep their job (Lynas, 2014a).

In January 2013, he delivered a public lecture to the Oxford Farming Conference which had an outstanding effect. In 2016, his website had attracted 187 comments, and he has become a celebrity activist and is now presented as a 'science advocate' and 'environmentalist writer' (Nebraska Today, 2016).

He began his 2013 lecture with apologies for 'ripping up GM crops', starting 'the anti-GM movement back in the mid-1990s' and 'assisting in demonising an important technological option which can be used to benefit the environment' (Lynas, 2013). He explained that his initial thoughts of GMOs were 'the stuff of nightmares': 'Mixing genes between species seemed to be about as unnatural as you can get – here was humankind acquiring too much technological power; something was bound to go horribly wrong. These genes would spread like some kind of living pollution' (Lynas, 2013). What made Lynas change his view on GM crops was his reflection on modern science. When he started to work on his book about global warming, he wanted to 'make it scientifically credible rather than just a collection of anecdotes'; he understood that there were 'people whom [he] considered to be incorrigibly anti-science' and he did not want to be like them (Lynas, 2013). Further, he undertook more reading on genetic modification and found that his 'cherished beliefs about GM turned out to be little more than green myths' (Lynas, 2013). With a new understanding, he is convinced that GM crops can help humanity to grow more food with less resources and that organic agriculture which is often portrayed as 'the naturalistic fallacy' (Lynas, 2013).

His interest in science brought him closer to scientists, and he even joined a university as a visiting fellow in Cornell University. He was then approached by the Bill and Melinda Gates Foundation which sent him to give public talks in Africa (Lynas, 2014a). He was prepared for the confrontation he would receive: the GM position is like swimming against the tide in that it can be 'ideologically suspicious'. He also points out that he regrets losing his friends (Lynas, 2014a).

So Lynas still remains part of the confrontation, but he has changed sides. He does not agree with his former colleagues anymore: 'These environmentalists picked up radical ideologies. . . . I can't bear people talking bullshit. To get an emotional reaction. It is a battle for science' (Lynas, 2014a).

He recognises that there is 'a clash of values' and criticises the people who had previously inspired him:

> West imperialism replicates. Vandana Shiva is against the development, opposition to modern world. Western imperial science is opposed to indigenous.

Stray cultish environmentalism produced the myth of romanticized poverty. It is about deep value level, can't be addressed with arguments. It is more symbolic than practical, such as 'genetic pollution'.

(Lynas, 2014a)

After his lecture and coming out, he participated in the debates but now on a different side. He supports pro-GM crop NGOs such as the Gates Foundation and the African Agricultural Technology Foundation (AATF) and meets with scientists, agronomists, farmers and policymakers. He also supported the Bt brinjal pilot project and the campaign behind the adoption of the crop, which became the first commercialised modified food crop in Bangladesh (Lynas, 2014b). He also argued in favour of Golden Rice in a talk entitled 'It's the 21st century, where's my GM rice?' (Lynas, 2014c). He now debates against anti-GM activists, raising doubts about their motives for participating in the debate. He recalled how he was approached by an anti-GM activist in Uganda: 'Straight after their presentation in the Parliament, came a person and told myths about cancer and fertility in connection to GM'. It was then covered in the *Uganda Observer* (Lynas, 2014a). He also suggested that the funding of such activism be investigated:

It is important to map the funding links. For anti-GM campaigns, activists would receive enveloped cash to carry out illegal authorities. By MP, a lot of powerful people. For example, Ghana Food Sovereignty is funded by Green Cult (Kult) in Belgium.

(Lynas, 2014a)

The 'Science Guy' and the Monsanto debate

William Sanford Nye is an American scientist, journalist and writer. He studied mechanical engineering in Cornell University where he also later taught. He also worked for Boeing and developed several patented inventions, including a small sundial that was included in the missions of the Mars Exploration Rover (Famous Scientists, 2016). In the early 1990s, he appeared in segments of the animated *Back to the Future* and got his own TV show, *Bill Nye, the Science Guy* from 1993–1998, which was promoted as a science education programme (Famous Scientists, 2016).

He is the author of several articles and books about science. In 2014, he wrote *Undeniable: Evolution and the Science of Creation* about the 'phenomenon of evolution'. In this he addressed the issue of GM food. While he discussed the potential benefits of such products (the possibility of a better taste), he raised his concerns against 'organisms designed in a top-down fashion' and human inputs in 'natural systems' which are already 'too complicated (Nye, 2015a, p. 235).

However, in 2015, he changed his views on GM after attending an anti-GM rally. What put him off was a conspiracy theory that the president of the US was controllable by Monsanto (Nye, 2015a, b). In his book, *Unstoppable: Harnessing*

Science to Change the World, he acknowledged that genetic modification is 'important for feeding humanity' (Nye, 2016, p. 230).

In his radio programme, as well as the book, Nye accepted that he had previously written against GM crops because 'it's always better to be cautious: you don't know what you gonna do to the ecosystem'. But he also accepted that since then he had conducted more research and had spent more time with scientists at the Boyce Thompson Institute for Plant Research at Cornell University watching scientists sequencing genes (Orwig, 2015). In addition, he visited the winner of the World Food Prize – 'the Nobel Prize for farming for agriculture' – Dr Robert Fraley, who also happens to work for Monsanto. In Nye's opinion, Fraley 'really is not such a bad guy and he believes that we can raise more food than ever on less land'. 'In other words we have 7.2, almost 7.3 blank people on 2% less land' (Nye, 2015b). Nye explained to his listeners the mechanism of applying glyphosate to fight against weeds and the benefits of RoundUp weed killer – the commercial name of glyphosate.

He also compared transgenic agriculture with organic farming and argued that GM food was safe and nutritious:

> There is a big difference in inputs from an agricultural standpoint. Organic agriculture takes a lot more water, and a lot more tillage. Actually you end up with less diversity of microbes in the soil, with modern RoundUp ready crops you don't have to till, to turn the soil however . . . [Y]ou can test the effects of the food . . . it has no effect on it as it's scientifically provable.
>
> (Nye, 2015b)

He acknowledged that RoundUp weed killer is not effective against one type of weed (pigweed), but it did not stop him from stating that RoundUp and GM crops are beneficial (Nye, 2015b). After Nye switched sides, Dr Fraley, obviously pleased with the outcome, in his article to the *Huffington Post* praised him for that act, arguing that 'the real sin is not in making a mistake, but in refusing to acknowledge it' (Fraley, 2016, para. 7). Nye was aware that his shift of ideas on GM crops would make media headlines, and he publicly reconfirmed his new views on GMOs (Nye, 2015b). He also received questions from the audience about his decision and had to deal with allegations that he was paid by Monsanto to do that. He declared that he was not paid by Monsanto and, describing his visits to Monsanto, he joked that he had dinner with the 'Monsantonian' team and paid for himself; yet he visited the Monsanto research centre and was offered and accepted a sandwich and a cup of coffee. He also tried to restore the public image of the company by denying the conspiracy theory of the 'thoughtless and short-sighted people' from the anti-GMO rally, stressing that Monsanto is not in the Top 5 companies, that it is, in fact, smaller than Apple. He argued that Monsanto does good for humanity by 'keep[ing] up in the race' against pests and weeds. He suggests that rejection of the technology is ignorance stance which is held by those afraid of new science (Nye, 2015b).

Farmers changing sides

Another participant group in the debate is that of farmers. Both proponents and opponents choose farmers as their constituency. Generally, there is a diversity of opinions among farmers, both in developed and developing countries. European farmers, as presented by farmers' unions and organisations, such as CP led by Jose Bove, appear to be rejecting GM crops. Yet there are farmers who grow GM crops in Europe, for example, in Spain.

A publicly recorded case of a European farmer who had been against GM crops and then changed his mind is that of Paul Temple. Mr Temple is a farmer involved in a family farming partnership and runs a 312-hectare farm, concentrating on beef cattle, cereals, oilseeds and vining peas in Driffield, East Yorkshire. In the past, he served as NFU vice-president and chairman of the Committee of Professional Agricultural Organisations (COPA/COGECA), an EU farmers' and agri-cooperatives lobby organisation based in Brussels. He founded the European Biotech Forum and from 2013–2015 was the Cereals & Oilseeds sector board member on the Agriculture and Horticulture Development Board (AHDB), a Levy Board which represents the farming of cattle, sheep, pigs, milk, potatoes, cereals, oilseeds and horticulture in the UK. He currently sits on National Non-food Crops Centre Board in the UK (N8 AgriFood, 2016).

The farm run by Temple and his family participated in the GM Field Scale Evaluation trials and is part of the Higher Level Stewardship scheme (N8 Agri-Food, 2016). He is also involved in CropLife International, a global network of the plant science industry represented by national associations (CropLife, 2015). That is, CropLife is an NGO and as it publishes articles about news in biotechnology development and the 'proven benefits of GM', it stands on the pro-GM side of the debate. In 2015, CropLife published an interview with Paul Temple, introducing him as 'a farmer who changed his mind on GM crops' (CropLife, 2015). In the interview, Temple acknowledged that he had 'lobbied to keep GM crops out of Europe in the 1990s' and that he 'thought it would help [UK] competitiveness by being GM-free'. However, he wanted to receive more information to decide on the use of GM crops and even took part in a field scale evaluation of GM crops in the period 1999–2001 (CropLife, 2015, para. 2). Explaining how he changed his mind, he said that it was his personal experience of studying GM crops that brought about this change of position:

> Seeing the benefits first hand opened my mind to the new breeding techniques. We could leave weeds within the crop over winter to benefit wildlife, the herbicide was easier to use and less harsh on the crop and our yields increased. We could tolerate weeds knowing we would be able to control them properly and precisely post-emergence. After three years I felt properly informed and offered to share my practical experience about the benefits of the technology.
>
> (CropLife, 2015, para. 3)

On the question of whether he would grow GM crops, he answered: 'But without a shadow of a doubt I would use GM crops if I could' (CropLife, 2015, para. 4). He can't grow them in UK at the moment due to the European restrictions on the commercialisation of GM crops. Thus, most of Temple's latest activity in the media is about lobbying for GM crops in Europe and the UK. Yet he remains pessimistic about the foreseeable future for GM crops in Europe because the European Commission plans to allow member states to opt out of allowing GM crops even if they have been approved at European level. He admitted that he feels the decision is 'a step back'. To gain personal experience, he travelled to Spain (the pro-GM EU member state) and compared the capacity of Bt maize crops and non-GM crops in the defense against the European corn borer (Driver, 2015).

In developing countries where GM crops are allowed, farmers also face the choice of whether to grow GM crops. Some, such as K. Ravichandran, the farmer with whom I met in India, have chosen to grow GM crops. Ravichandran has been involved in farming for the last 30 years and inherited his farm in a village Poongulam, Tamil Nadu. Because he manages the land he inherited along with plots belonging to his sisters, he has a total of 60 hectares of land, making him by Indian standards a big farmer. He grows rice, sugar cane, coconut and Bt cotton. The decision to grow Bt cotton was made after running a small experiment on three plots of land on which he planted transgenic, conventional and organic seeds and then compared his yields and costs. As a result of his experiment based on cost benefit approach, he switched to the GM variety (Ravichandran, 2015).

In Burkina Faso, the government revised the regulatory system on bio-safety, and after signing a joint venture with Monsanto in 2008, the country became the tenth in the world to grow Bt cotton commercially (Savadogo, 2014). This then became a significant argument in favour of GM crops. The anti-GM NGOs, both international and national Burkinabe ones such as GMWatch and the National Union of Agropastoral Workers (Syntapa), responded to the fact that Bt cotton fields accounted for 70% of all Burkinabe cotton production with an enforced critique of GM crops, arguing that 'Monsanto and other biotechnology firms will use Burkina as a Trojan Horse to spread GMOs in the subregion' (Combat Monsanto, 2011).

But in 2016, the Burkinabe President, Roch Marc Christian Kabore, announced the decision of the government to withdraw the permission to grow GM cotton from 2018. The reason for such a decision is that 'farmers had complained that genetically modified cotton yield shorter fibers than the conventional variety and it was no longer possible to manufacture the smooth and stable thread which was essential for textile production' (DW, 2016, para. 3). This raised the dispute with Monsanto as farmers have claimed compensation from Monsanto at an estimated €74 million (DW, 2016). The case of Burkina Faso possibly changing its stand in relation to the GM debate might have serious implications for the debate and which side will gain more support from the global community. An NGO that supports GM and promotes science-based ideology, Genetic Literacy Project, published an article by a member of the Cornell Alliance for Science which tried to cool down the allegations of Burkina Faso abandoning Bt cotton, explaining that 'reports of GM cotton's death are also an exaggeration'. It further stated that Burkinabe researchers are working together with Monsanto to introduce the insect-resistant trait into local long-fiber cotton varieties (Conrow, 2016).

In their turn, anti-GM NGOs used the case of Burkina Faso in anti-GM propaganda. For example, Friends of the Earth Nigeria claimed that 5 million Nigerian farmers and civil society members had signed a petition listing their concerns about GM crops and hoping to avoid the experience of their neighbours (DW, 2016).

Oxfam and WWF: withdrawn position

The position of two international NGOs, Oxfam and the World Wildlife Fund (WWF), can be referred to as having been withdrawn, and the experiences of such organisations in making and presenting arguments about GM crops offer an example of how politicised the topic is and how much pressure can be laid upon debate participants. Oxfam is an NGO focusing on issues of international development, while WWF is an international environmental NGO. The American branches of both organisations made claims about the possible benefits of GM crops using various arguments based on social development and environmental protection. The reaction of partnering organisations and supporters forced both to withdraw these earlier claims and issue official statements emphasising that they do not support GM crops.

While in the 1980s to the early 1990s, Oxfam-UK was clearly opposed to GM crops (Myers et al., 1992), its American arm had commissioned research on Bt cotton, a GM crop, in the 2000s, which drew cautious conclusions that GM crops could have some benefits.

Chapter 5 has discussed the Oxfam study, 'Biotechnology and Agricultural Development', which summarised the results from the Oxfam-America project entitled 'Learning from the Experience of Small-Scale Farmers: The Case of Transgenic Cotton'. It studied the experience of growing GM cotton in developing countries and discussed 'the relevance of agricultural biotechnology for resource-poor farmers' (Tripp, 2009, p. 1). The authors admitted that they had 'to examine a very complex and controversial subject' and did not insist on the universality of their findings due to research methodology limitations. However, the data found proved the decrease in insecticide costs and increased yields resulting from the adoption of Bt cotton occurred in a number of countries including China, India and South Africa (Tripp, 2009, p. 74). Despite such cautiousness in introducing the study, the organisation has been heavily criticised by the anti-GM activist organisations such as GMWatch and GRAIN.

GMWatch published an open letter to Oxfam USA addressing the publication of the research. The letter is signed by six NGOs: African Centre on Biodiversity, Indian Farmers Association Bharatiya Krishak Samaj, Center for Food Safety, National Coordination of Peasants' Organizations (Mali), Farmers Association of Tamil Nadu and the Oakland Institute. The letter argued that with the publication of the project on Bt cotton, Oxfam appeared to be 'a good broker' for Bt cotton promotion in Africa (GMWatch, 2010). It made allegations that there could be some support involved from the Gates Foundation, a well-known supporter of transgenic agriculture. While there is almost nothing wrong with that, GMWatch reminds Oxfam USA leaders about Oxfam's mission to support poor people and their livelihoods and explains that it is 'deeply troubled that the study and its scientifically questionable (at best) conclusions, falsely support practices that hinder rather than help efforts to save lives, end poverty, and promote social justice' (GMWatch, 2010). Accusations regarding conducting false research and supporting corporate interests ('Oxfam America is surrendering to the biotech industry and their corporate extensions and private foundations') are very serious:

The publication betrays the vibrant global movement that is demanding a more ecologically sustainable and socially just agriculture, free from corporate control. In reviewing the publication we find it problematic for the following reasons, which we elaborate upon in this letter:

1. False advertising on appearing neutral while endorsing GM crops
2. Incomplete research using selective information to arrive at a pro-GM conclusion
3. Its focus on GM crops as a solution to help resource-poor and subsistence farmers climb out of poverty

(GMWatch, 2010).

In London, I met with a former employee of Oxfam America who shared their views on what has happened. Originally an opponent of GM crops, this person, who for obvious reasons would want to remain anonymous, has been working on agricultural projects in Oxfam America in developing countries. Having studied seed markets, the informant came to know the challenges of modern smallhold farmers, such as lack of education, regulatory constraints, issues of property rights and climate change. After reading a book by Robert Paarlberg (*Starved for Science: How Biotechnology Is Being Kept Out of Africa*), their viewpoint has changed on the use of GM technology in agriculture. The book 'opened their eyes' and made them 'angry' about 'outrageously misleading in facts'. This person was not the only one with such ideas on GM crops in Oxfam. At the same time, other branches of Oxfam, particularly Oxfam Germany, were 'very anti-GM' and 'the global organization struggled to come up with a strategy defining its position' (GMWatch, 2010). The organisation held different meetings with a number of NGOs involved in the debate, most of which 'trashed in a discussion about farming in the USA', accusing Oxfam of 'giving voice to Monsanto', and 'empowering Monsanto'. The organisation was forced to apologise and the interviewee 'personally got attached' and finally left their job (GMWatch, 2010).

In 2011, Oxfam International signed off an official position on GM crops denying its support to GM crops, 'to clear . . . top lines'. The Advocacy and Campaign Director proposed the statement, yet even in this announcement Oxfam pointed out that 'Oxfam affiliates as part of Oxfam International have not to date reached consensus on a strong Oxfam International policy position clearly for or against GMOs' (Oxfam, 2011, p. 1).

It states that Oxfam, 'as a rights-based international development organiza-tion', argues that 'decision-making on the use of genetically engineered organisms should be based on how they impact on human rights, which can be translated into the principles of participation, transparency, choice, sustainability and fair-ness'. While Oxfam recognised that 'GMOs represent a good business for some, there is a mention that for poor farmers the benefits and relevance of GM crops are 'questionable' (Oxfam, 2011, pp. 1–2).

Technically, the official position of Oxfam is anti-GM but because it has reviewed its claims about the possible benefits of GM crops, it is an example of an organisation that has withdrawn its views. Both cases illustrate the geographic factor in the assessment of GM crops: the American branches were more willing to put forward arguments in favour of their use than its branches in Europe. It is also an illustration of how the views and treatments of GM can vary within the same organisation.

According to Mark Lynas who was engaged in the project of Oxfam's 'Growth Camp' on climate resilient agriculture, 'the organization is internally divided', and 'Oxfam is against GM, because it is very anti-corporations'. But there is an internal divide between North and South. Interestingly, his colleagues – members of the UK branch – are not vivid opponents of GM and avoid much engagement in the debate, so 'Oxfam GB is alright' (for GM crops). He has mentioned that behind the scenes, Oxfam is holding internal talks. Once they can see that 'all boxes are ticked, such as that peasants are having control in the process and benefits are so clear and risks are locked down', they will be 'more open' in presenting their position (Lynas, 2014a).

He also mentioned that similar processes of internal divide in GM crop views might exist in another global NGO – Friends of the Earth, a traditional opponent of GM crops: 'Friends of the Earth International are shouting imperialism, but the FoE UK removed all papers from their webpage' (Lynas, 2014a).

As in the case of the Oxfam project on Bt cotton, GMWatch also published an online critique of another international NGO – WWF – criticising it for its position on GM crops, this time for GM soya.

WWF's entry in the debates on GM crops began with a report by WWF Switzerland 'Genetically Modified Organisms (GMOs): A Danger to Sustainable Agriculture. It was written by economist Gerald Assouline and sociologist Tereza Stöckelová and pledged for a moratorium of GM crops. The report was updated in 2005, probably in order to address another point of view on GM crops put forward by WWF USA (WWF, 2005). The report claimed that there were no economic benefits from growing GM crops and no scientific evidence on safety of such crops (WWF, 2005).

WWF USA started a foundation of the Roundtable on Sustainable Soy in 2004 and then renamed it to Roundtable on Responsible Soy (RTRS). It was criticised for including agro-corporations which included Monsanto, Syngenta, and Cargill in its RTRS. The roundtable was called 'a mega-green washing operation . . . to cover up the environmental and social destruction caused by soy production in South America' (Rulli, 2010). WWF was accused of ignoring human rights violations by agribusiness and having connections in the WTO, and its schemes of environmental certification were called 'meaningless'.

All three accusations are very sensitive since they tackle the exact values and campaigns that the organisation used to promote.

Over 60 organisations sent an open letter to RTRS members calling for its abandonment. WWF officially responded that the initiative was 'technology neutral' and that 'if the RTRS principles and criteria included a prohibition on the use of GMs, their potential application would be restricted to the limited proportion of global production (estimated at 30%) that is GM-free', thus 'limit[ing] the potential of the RTRS to address impacts of GM soy production as well' (Roth, 2009).

The critique targeted Jason Clay, the vice president of WWF USA, portraying him as 'an enthusiast of market-based mechanisms to tackle environmental problems and a fervent promoter of agrofuels'. There is nothing wrong with that and it might be a good picture of Clay's views, however, such media attention resulted in some WWF supporters from the environmental movement withdrawing their support, including financial support.

In 2009, Jason Clay of WWF USA explained that the organisation felt that all genetic work could be useful, and technology can be neutral if there is focus on measurable results in any agricultural production system (Clay, 2010). In 2012, WWF issued a position statement on GMOs, emphasising that 'WWF does not promote or endorse the use of GMOs; applies a precautionary approach to the introduction of GMOs; and advocated the retention of non-GMO options for all relevant commodities'. In its background notes, WWF stated that 'technological advances can help create benefits for the environment and society, but the benefits have to be clearly demonstrated and the risks avoided and/or limited'. It also noted that under the condition of the absence of international consensus on the risks or benefits from the use of GM crops, such assessment is 'subject not only to scientific interpretation, but also to differences depending on environmental, social and economic contexts' (WWF, 2012).

The GM soya story returned to public attention when Wilfried Huismann began a journalistic investigation. In June 2011, he produced a documentary titled "The Silence of the Pandas: What the WWF Isn't Saying". It was later published as a book on internal dynamics in 'the green empire' called *WWF PandaLeaks: The Dark Side of the WWF* and he 'found [a] few skeletons in the closet' (Huismann, 2014). GM soya was one of them. While Huismann acknowledged that most European branch representatives rejected GMOs and considered the whole topic to cause much pain, there were influential GM soya supporters in the Americas. In addition to Jason Clay, a former leader of WWF Argentina, Dr Hector Laurence, was another GM soya supporter who openly argued that it was 'greener' and beneficial to the protection of the environment since it helped to reduce the use of chemicals and tillage (Huismann, 2014, pp. 191, 197). Huismann discovered that previously Dr Laurence had served as a representative of two agrochemical big businesses: Morgan Seeds and Pioneer, a subsidiary of DuPont. Dr Laurence was bold enough to tell Huismann, a German who was most possibly a vivid opponent to GMO:

> You Europeans must be told straight out that you are quite backward in some areas, especially when it comes to modern technologies. You have become the victims of the leftwing hysterics who denounce genetic engineering as the work of the devil – instead of listening to science.
>
> (Huismann, 2014, pp. 180–181)

According to Lars Langenau, another German journalist, the reputational risks from Huismann's book were taken and addressed very seriously by WWF management: it made every effort to prevent sales of the book through its lawyers and readers could only get the book online (Langenau, 2012). After such a scandal, it is hardly surprising that WWF has not participated in the GM crop debate since then.

Literature

AllowGoldenRiceNow (2016) Home webpage. www.allowgoldenricenow.org/ as viewed 28.09.2016.

Baulcombe, D. (2014) Interview. Cambridge, 6 February 2014.

Clay, J.W. (2010) *Agriculture from 2000 to 2050—the business as usual scenario.* Washington, DC: The Global Harvest Initiative.

Combat Monsanto (2011) Burkina Faso is a Trojan Horse for GMOs in Africa interview with Ousmane Tiendrébéogo, Secretary General of the National Union of Agropastoral Workers (Syntapa). *Journal of Alternatives*, 28 June 2011. http://journal.alternatives.ca/spip.php?article6231

Connor, S. (2014) Former Greenpeace leading light condemns them for opposing GM 'golden rice' crop that could save two million children from starvation per year. *The Independent*, Thursday 30 February 2014. www.independent.co.uk/news/science/former-greenpeace-leading-light-condemns-them-for-opposing-gm-golden-rice-crop-that-could-save-two-9097170.html

Conrow, J. (2016) Burkina Faso putting GMO cotton on hold, not abandoning it. Genetic Literacy Project, 29 April 2016. www.geneticliteracyproject.org/2016/04/29/burkina-faso-putting-gmo-cotton-on-hold-not-abandoning-it/ as viewed 28.09.2016.

Cook, G., Pieri, E., Robbins, P. (2004) The scientists think and the public feels: Expert perceptions of the discourse of GM food. *Discourse and Society*, 15(4): 433–449.

CropLife (2015) Why this farmer changed his mind on GM crops. 4 March 2015. https://croplife.org/news/why-this-farmer-changed-his-mind-on-gm-crops/ as viewed 28.09.2016.

Dobermann, A. (2015) The future directions of Rothamsted Research. Plant Science Seminar, Cambridge. 22 January 2015.

Driver, A. (2015) No GM approval in foreseeable future, despite changing attitudes: Temple. *FG Insight*, 15 June 2015. www.fginsight.com/news/no-gm-approvals-in-foreseeable-future-despite-changing-attitudes – temple-4094 as viewed 28.09.2016.

DW (2016) Burkina Faso abandons GM cotton. www.dw.com/en/burkina-faso-abandons-gm-cotton/a-19362330 as viewed 28.09.2016.

Famous Scientists (2016) Bill Nye. www.famousscientists.org/bill-nye/ as viewed 28.09.2016.

Fraley, R.T. (2016) Bill Nye's change of heart on GMOs is in the best scientific tradition. 26 January 2016. www.huffingtonpost.com/dr-robert-t-fraley/bill-nyes-change-of-heart_b_9055296.html as viewed 28.09.2016.

Gaudilliere, J.P. (2006) Globalisation and regulation in the biotech world: The transatlantic debates over cancer genes and genetically modified crops, in *Global power knowledge science and technology in international affairs. OSIRIS*, 21: 251–272.

GM-Free (1999) Keeping your life and environment free from genetically modified food. 1(2), June–July 1999. Skelmersdale: KHI Publications.

GMWatch (2010) Open Letter to Oxfam America. www.gmwatch.org/en/news/archive/2010/12130-open-letter-to-oxfam-america as viewed 05.05.2017.

Greenpeace (2013) Golden Illusions: The broken promises of Golden Rice. www.greenpeace.org/international/en/campaigns/agriculture/problem/Greenpeace-and-Golden-Rice/ as accessed 12.06.2017.

Haerlin, B. (2014) Interview. Berlin, 8 March 2014.

Harrison-Dunn, A.R. (2014) What do the eco-heretics mean for GM golden rice? Nutra Ingredients. www.nutraingredients.com/Consumer-Trends/What-do-the-eco-heretics-mean-for-GMgolden-rice as viewed 01.06.2014.

Huismann, W. (2014) *PandaLeaks: The dark side of the WWF*. Bremen: Nordbook.

International Criminal Court (2002) Rome Statute of the International Criminal Court. The Text circulated as document A/CONF.183/9. The Hague: ICC. www.icc-cpi.int/nr/rdonlyres/ea9aeff7-5752-4f84-be94-0a655eb30e16/0/rome_statute_english.pdf as viewed 01.06.2014.

Juncker, J.C. (2014) *A new start for Europe: My agenda for jobs, growth, fairness and democratic change*. Political Guidelines for the Next European Commission. Opening Statement at European Parliament Plenary Session. Strasbourg: European Parliament.

King, A., Rautner, M., Tyler, G. (2011) *Golden rice's lack of lustre*. Amsterdam: Greenpeace International.

Kuklinsky, J.H., Quirk, P.J. (2012) Reconsidering the rational public: Cognition, heuristics, and mass opinion, in A. Lupia, M.D. McCubbins (eds.) *Elements of reason cognition, choice, and the bounds of rationality*. Cambridge: Cambridge University Press, pp. 153–182.

Langenau, L. (2012) Die Dunkle seite des Panda. *Sueddeutsche Zeitung*, 28 May 2012. www.sueddeutsche.de/wissen/schwarzbuch-wwf-die-dunkle-seite-des-panda-1.1366518

Lomborg, B. (2013) Meaning good, doing bad. *The European*, 2013(9). www.theeuropean-magazine.com/bjorn-lomborg/7481-rejection-of-gm-rice

Lovelock, J. (2009) *The vanishing face of Gaia: A final warning*. London: Allen Lane.

Lovelock, J. (2012) Carey and Wolfe Valley opposition to wind turbines. Email from 12 December 2012. www.bishop-hill.net/storage/James%20Lovelock%20Letter.pdf as viewed 28.05.2014.

Lupia, A., McCubbins, M.D. (2012) The institutional foundations of political competence: How citizens learn what they need to know, in A. Lupia, M.D. McCubbins (eds.) *Elements of reason cognition, choice, and the bounds of rationality*. Cambridge: Cambridge University Press, pp. 47–66.

Lynas, M. (2008) *Six degrees: Our future on a hotter planet*. London: Harper Perennial.

Lynas, M. (2011a) *God species: How the planet can survive the age of humans*. London: Fourth Estate.

Lynas, M. (2011b) Why I will never be an 'ambassador' for the corporate biotech lobby. Entry to the website. www.marklynas.org/2011/10/why-i-will-never-be-an-ambassador-for-the-corporate-biotech-lobby/#sthash.ubXuieBF.dpuf as viewed 01.06.2014.

Lynas, M. (2013) Lecture to Oxford Farming Conference. Oxford, 3 January 2013. www.
    marklynas.org/2013/13/01lecture-to-oxford-farming-conference-3-january-2013
Lynas, M. (2014a) Interview with Mark Lynas. Oxford, 8 February 2014.
Lynas, M. (2014b) Debate: What's wrong with GM? Why I turned from GM opponent to
    advocate. www.scidev.net/global/biotechnology/opinion/gm-opponent-to-advocate.
    html as viewed 28.09.2016.
Lynas, M. (2014c) Plenary speech to International Rice Congress. 31 October 2014. www.
    marklynas.org/2014/10/mark-lynas-plenary-speech-for-international-rice-congress-
    2014-bangkok-thailand/ as viewed 28.09.2016.
Marleau, R., Montpetit, C. (2000) *House of Commons procedure and practice.* Montreal:
    Canada Parliament, House of Commons.
Monbiot, G. (2002) The Covert Biotech War. *The Guardian*, 19 November 2002. www.
    theguardian.com/science/2002/nov/19/gm.food
Monbiot, G. (2010) Deep peace in Techno-Utopia. Blog Entrance from 5 November 2010.
    www.monbiot.com/2010/11/05/deep-peace-in-techno-utopia as viewed 03.03.2014.
Moore, M. (2014) Interview with Michael Moore. London, 31 January 2014.
Moore, P. (2014) Golden rice now: Preventing it is a crime against humanity. www.
    allowgoldenricenow.org as viewed 25.05.2014.
Myers, D., Davidson, J., Chakraborty, M. (1992) *No time to waste: Poverty and the global
    environment.* Oxford: Oxfam.
N8AgriFood (2016) Paul Temple. http://n8agrifood.ac.uk/speaker/paul-temple/ as viewed
    28.09.2016.
Nebraska Today (2016) Science advocate Mark Lynas to present Heuermann Lecture.
    http://news.unl.edu/newsrooms/today/article/science-advocate-mark-lynas-to-present-
    heuermann-lecture/ as viewed 28.09.2016.
Nye, B. (2015a) *Undeniable: Evolution: Science creation.* New York: St. Martin's Griffin.
Nye, B. (2015b) *Cosmic queries: GMOs with Bill Nye.* Star Talk Radio Show. Season 6,
    Episode 26. www.startalkradio.net/show/cosmic-queries-gmos-with-bill-nye-part-1/ as
    viewed 28.09.2016.
Nye, B. (2016) *Unstoppable: Harnessing science to change the world.* New York: St. Martin's
    Griffin.
Orwig, J. (2015) What changed Bill Nye's mind on GMOs? *Business Insider*, 25 November
    2015. http://uk.businessinsider.com/bill-nyes-stance-on-gmos-2015-3?r=US&IR=T
Oxfam (2011) *Oxfam international position on GMOs.* London: Oxfam. Signed off March
    2011.
Potrykus, I. (2001) Potrykus responds to Greenpeace's criticism of 'golden rice'. AgBioView.
    Archive Message #979 from 9 February 2001. http://agbioview.listbot.com/cgi-bin/
    subscriber?Act=view_message as viewed 01.06.2014.
Ravichandran, K. (2015) Interview with K. Ravichandran. Bangalore, 26 February.
Roth, S. (2009) WWF and Monsanto: Is GM Soy now okay? *The Ecologist*, 19 June 2009.
    www.theecologist.org/News/news_analysis/271944/wwf_and_monsanto_is_gm_soy_now_
    okay.html
Rulli, J. (2010) *World Wildlife Fund: Loyal ally of corporate agribusiness and Monsanto.*
    Organic Consumers Association. www.organicconsumers.org/news/world-wildlife-fund-
    loyal-ally-corporate-agribusiness-and-Monsanto
Sahai, S. (2004) Golden rice: Not food for the poor: New developments in Syngenta and
    Humanitarian Board. *Economic and Political Weekly*, 39(42): 4612–4613.
Savadogo, M. (2014) Case study: Burkina Faso, in D.P. Keetch, D. Makinde, C.K.
    Weebadde, K.M. Maredia (eds.) *Biosafety in Africa: Experiences and best practices.* East
    Lansing: Michigan University Press, pp. 82–88.

Sniderman, P.M., Brody, R.A., Tetlock, P.E. (1991) *Reasoning and choice: Explorations in political psychology*. Cambridge: Cambridge University Press.

Tripp, R. (ed.) (2009) *Biotechnology and agricultural development: Transgenic cotton, rural institutions and poor-resource farmers*. London: Routledge.

Vidal, J. (2014) WWF International accused of 'selling its soul' to corporations. *The Guardian*, 4 October 2014. www.theguardian.com/environment/2014/oct/04/WWF-international-selling-its-soul-corporations as viewed 10.10.2016.

Vrain, T. (2013) Former pro-GMO scientist speaks out on the real dangers of genetically engineered food. *PreventDisease.Com*. http://preventdisease.com/news/13/050613_Former-Pro-GMO-Scientist-Speaks-Out-On-The-Real-Dangers-of-Genetically-Engineered-Food.shtml as viewed 29.05.2015.

WWF (2012) *WWF position statement: WWF's position on genetically modified organisms (GMOs)*. Gland: WWF.

Zelko, F. (2013) *Make it a green peace!: The rise of countercultural environmentalism*. Oxford: Oxford University Press.

10 The global transfer of ideas

This chapter discusses how the spread of ideas happens across national boundaries in the GM debate. The debates on GM crops are a global phenomenon, covering all major continents, except Antarctica. What is peculiar is that while there are regional variations and political geographic patterns, such as Europe being known for its opposition to GM crops and North America being mostly supportive of GM agriculture, similarities exist in how the debates are processed in different regions. There is also a pattern of global diffusion of ideas. Members of NGOs in some countries affect representatives of NGOs and policymakers in other countries and transfer their rhetoric across national borders. It also seems to be the case that activists learn successful practices from their counterparts in international activism and there are similar worldwide responses by public and private society to such activism. This spillover of ideas is referred to as a global transfer or diffusion of ideas.

The common saying is that revolution begins in people's minds. So then we just need to create an idea backed up by selected reasons and interests; next, it should be spread among other individuals, allowing the group to promote the idea. Being part of the global network, through developing contacts, reaching out to NGOs and helping to create new NGOs, activists within the GM crop debate are able to influence the debate in countries outside their homeland. The alterglobalist activist Alex Callinicos once said in reflection on how a social movement works:

> One of the great beauties of our movement – and the forums that have emerged from and helped to sustain it – is the way in which people from all sorts of backgrounds and with the most diverse preoccupations come and mix together, participating in a process of mutual contamination in which we learn and gain confidence from one another.
>
> (Maeckelbergh, 2009, p. 6)

This interesting use of the word 'contamination' reflects how all sides interact; exchange their ideas, values, interests; and either borrow from each other or create new joint ones in the process of interaction. In formal institutions such communication is also possible, as proven by the case of European NGOs lobbying European institutions.

Another insight on how groups are formed through the influence of external ideas is from Malcom Gladwell, who has shown the mechanism for such transfer and conditions necessary for an idea to become epidemic. What he is doing is referred to as 'the biography of idea', explaining the emergence of fashion shows, changes in criminal dynamics, the sudden success of common items like Hush Puppies shoes and the phenomenon of word of mouth. In fact, he is showing real examples of how to create a grassroots movement successfully: he describes the evolution of the campaign towards awareness of diabetes and breast cancer in the black community of San Diego (Gladwell, 2009). He also argues that ideas and consequently the products of such ideas spread as widely as epidemics or viruses. Arguably, debates on GM crops are another example of how ideas, particularly those against them, became viral within a global network of NGOs.

European influence in African debates over GM crops

It is a matter of fact that genetic research began in Europe and the region has greatly contributed to genetics: Charles Darwin, Gregor Mendel, Friedrich Miescher and Archibald Garrod and Francis Crick were the Europeans who laid out the important foundations for contemporary gene research. They announced about their discovery in the pub called Eagle located next to King's College, Cambridge. But why has Europe been anti-GM in the recent decades?

To understand why European policymakers and the public are anti-GM, several reasons can be examined. To compare Europe and the US, there are reverse trends in regulation on GMOs in the two regions: in the period from the 1960s to the late 1980s, the US had in place stricter regulation on risk in relation to health and the environment, while European regulators worked in cooperation with industry and NGOs were given little access to public policy. However, the European regime changed to become based on the precautionary principle and is now stricter than the American one (Lynch and Vogel, 2001). The strictness of the European regime was embedded within the local and institutional context (e.g. the BSE outbreak in the UK). In contrast to the US where several institutions were made responsible for GM regulations – the Environmental Protection Agency (EPA), United States Department of Agriculture (USDA) and Food and Drug Administration (FDA) – the major institutional upper hand in the European Union was given to the Directorate General on Environment, Consumer Protection and Nuclear Safety (DG XI) which took a more cautious approach on GM technology than other directorates and was able to win over the opposition from other directorates, such as the Directorate General on Science, Research and Development (DG XII) (Lynch and Vogel, 2001).

As a result, despite previously strict limits levied on GMOs in the past, scientists in the US work within more favourable conditions for the approval of field trials and patents, and the country has the highest acreage of biotechnology crops in the world (ISAAA, 2015).

An additional reason used by European policymakers in adopting strict control on GM crops is public opinion. EU surveys have revealed that scientists and pro-GM activists remain in the minority, while most Europeans are against the use of GM crops. Only 27% of Europeans indicated a positive attitude to GM food, according to a 2005 Eurobarometer poll, although this is still an increase in comparison with 21% in 2002 (Gaskell, 2006).

While national public opinions do vary from year to year, it is possible to identify that Austrians and Germans have the highest degree of mistrust in relation to GM technology: 41% and 33% respondents respectively in both countries identified the effects of biotechnology as negative in 2010. The German public has also claimed to be the most knowledgeable on GM foods, with 95% of the population having heard of GM technology (EC, 2010). While it may be possible to explain such suspicion and sensitivity of German-speaking countries regarding genetic research as being attributable to the painful legacy of the World War II when inhuman research trials were supported by the Nazis, other European countries such as those in Scandinavia have no obvious reason to avoid GM crops.

Paarlberg tried to explain the lack of motivation on the part of European farmers to grow the first GM crops designed in the 1990s by the fact that these were soy, maize and cotton and not the main European crop, so only 1% of farmers could benefit from growing them (Paarlberg, 2014, p. 174). However, as there is a high demand for the soybean in Europe, as shown in Chapter 4, Europe is still anti-GM with the partial exception of Spain, whose farmers do grow it. Thus, it does seem that there are other reasons contributing to the continuing opposition in Europe to GM crops.

Some economists, such as Lee Ann Jackson and Kym Anderson, have argued that in much more densely populated Europe the co-existence regimes of combining GM and conventional crops are difficult to implement, and the possible contamination from GM fields can bring high economic costs (Jackson and Anderson, 2005). For example, European authorities adopted the Novel Foods Regulation in 1997 which also covered GM food. However, it was 'largely inoperable' as there were no specific details provided regarding implementation and individual EU member states were left to define thresholds, testing methods, products for testing and the content of labels. This led national governments to adopt their own measures. Austria initiated the process, being first to ban GMOs in 1997. This was followed by Luxembourg, France, Greece and Germany, and in 1999, the Council of Environmental Ministers stalled the approval process until updated regulations were adopted (Heijden, 2010, p. 55). It is obvious that European member states influence one another. The case of the Séralini study, discussed earlier, has also shown that the influence of the European anti-GM lobby reached beyond EU borders.

That diversity in treating GM crops, with a slight preference to the anti-GM lobby, across Europe is still maintained by the political decisions of European leaders. For example, the president of the European Commission, Jean-Claude Juncker, argued for a review of the legislation on the authorisation of GMO in order 'to give the majority view of democratically elected governments at least the same weight as scientific advice, notably when it comes to the safety of the food we eat and the environment in which we live' (Juncker, 2014).

Thus, to be precise, the EU has adopted a mixed, almost double standard approach. While the European Commission has funded a number of environmental NGOs active in promoting anti-GM propaganda, there are two GM projects – AMIGA and GRACE – funded by the EU which are purposely kept 'in shadow' (Hoegel, 2014). All this allows diverse initiatives that support both sides of the debate within the EU, but it would appear confusing for those without specific knowledge.

The importance of the economic argument in the discussion of GM crop application should not be downplayed, but the role of civil society groups in Europe in influencing political processes and forming public opinion has been exceptionally high. It is based on the influence that NGOs are allowed to exert in European political processes and the complex nature of European institutions. Both Greenpeace and EuropaBio which led anti-GM and pro-GM campaigns have access to European policymakers in Brussels. In fact, both are located there.

European institutions involved in the decision-making process on GM crops were engaged in 'an Ancient Greek drama, full of ideological conflicts and power play between different Commission DGs, EP commissions and Councils and, gradually, with the officials on environment getting into the driver's seat' (van Schendelen, 2005, p. 181). The anti-GM lobby won the battle in the 1990s.

Two strategies used by civil society groups lobbying against GM can be identified in Europe: direct blockage and 'advocacy science' – discussed in Chapter 9. The first strategy is aimed at the physical destruction of crops. It has been adopted by a number of radical NGOs, particularly Greenpeace which first engaged in the GM debate in 1994. By the end of the century, the campaign had reached its momentum and was accompanied by much attention from media which picked up 'the problem with public consent' and applied the term 'Frankenstein food' (used in a *Daily Mail* newspaper headline). In 1997, Greenpeace protesters blocked the ship carrying genetically modified soybeans in Rotterdam, the Netherlands. In 2000, protesters blocked a ship off Anglesey, UK, and placed a banner with the words "Europe says no to GM" (BBC, 2000). On numerous occasions, protesters have destroyed trial crops: in the UK (2000), in France (2006, 2010), in Spain (2010) and later in 2011 in Germany, just to cite a few incidents. Such campaigns appear to be commonly organised all over the world but occur particularly in Europe although all were arranged by different, rather independent groups, since Greenpeace operates as a network (Parr, 2014). Such political activism of European NGOs contributes to the current European regulation.

However, what seems to be a matter of internal choice for Europe has direct international implications. Europe, as a participant in international politics, has been criticised for its stand on GMOs. In 2003, the US, jointly with Canada and Argentina, opened a case in the WTO for a dispute settlement panel demanding the European Union to justify its moratorium on GM imports. Robert Zoellick, the US trade representative in Africa, has even labelled European policymakers 'Luddites' and pinned the African rejection of GM food aid (particularly the case of Zambian hunger) on it (Weasel, 2009).

It is also interesting that while European pro-GM activists often cannot win the debate using the argument that Europe itself needs GM crops, they refer to the European influence on developing countries and criticise the anti-GM European lobby.

For example, Belgian Professor Baron Marc Van Montagu led the research on the transfer of genes between bacteria and plants and founded one of the early biotechnology companies in the 1980s, Plant Genetic Systems (PGS) – now part of Bauer – to commercialise insect-resistant and herbicide-tolerant crops. In 1998, he also co-founded Crop Design, a company that produces commercial GM corn and rice seeds for the global market. He experienced the opposition of Europe-based anti-GM crop NGOs firsthand when PGS was sued by Greenpeace. Green-peace demanded the retraction of the company's patent for genetic engineering in plants against fungi and weeds in 1992 'on the grounds that the grant of a patent for plant life forms and the exploitation of the patent was contrary to morality and/or "ordre public"' (Article 53(a) EPC cited in EPO, 1995).

When he retired, Baron Van Montagu became engaged in activism. In 2000, he founded the Institute of Plant Biotechnology Outreach (IPBO), a subsidiary of VIB University in Ghent, Belgium. As a non-profit, it is equivalent to an NGO. Its mission is to raise awareness about biotechnology and its potential solutions for sustainable agricultural and industrial systems, along with its ability to conduct 'a concerted effort' and 'converting this promise into practice in less developed regions' (VIB, 2013). In other words, the Institute and Van Montagu intend to counter-argue the existing anti-GM rhetoric in Europe. He is thus referred to as a representative of the European counter-movement on GM crops. A large part of his current activity is '[to] coach networks of public and private activities, so that knowledge acquired can be used for the innovations so badly needed' (Van Montagu, 2015, p. 325).

In a speech delivered at the 2013 World Food Prize event, equivalent to the Nobel Prize in the field, he discussed how neither he nor his fellow scientists had antici-pated the resistance to genetic modification in agriculture. He argued that genetic engineering in crops was equal to selective breeding mechanisms that have existed 'since the dawn of agriculture' (Van Montagu, 2013). He made the case for GM crops, offering a possible solution to the challenges presented by climate change, and placed the moral responsibility upon the opposition movement in Western countries for delaying the delivery of the technique to developing countries and small farmers (Van Montagu, 2013). In opposition to the alterglobalist argument, he claimed 'one irony' that the 'hyper-regulation of GM crops' caused by the opposition led to higher costs in getting GM seeds to the market, resulting in their being accessible to only large companies and research units in the West, but 'smaller enterprises in developing countries' cannot afford them (Van Montagu, 2013).

African leaders from the countries in which GM technology has already been commercialised have advised their counterparts in other African nations to prepare the regulatory base first (Pandor, 2014). Understanding the complexity involved may lead to a more careful approach to the reliance on GM crops in the region.

Some experts based in Africa follow Van Montagu's line of argument that the European opposition to GM crops results in negative outcomes in Africa's capac-ity to address food insecurity and the climate change challenge. For example, Dr Joe DeVries, an American citizen based in Africa since 1986 and director of AGRA's Programme for Africa's Seed Systems (PASS), has argued that high dependency on food aid to address hunger and food insecurity in Africa has been a negative factor. In the worst cases, large amounts of food aid may contrib-ute to increased conflicts over resources and political power, misuse and further inequality, resulting in genetic engineering being considered a way forward for Africa (DeVries, 2013).

The European communication to Africa regarding the use of GM has been mixed and complicated. On the one hand, European scientists have developed crops that are potentially beneficial for African farmers, such as Golden Rice (Moore, 2014). On the other, many European NGOs, such as Earth Open Source, GRAIN and GMWatch amongst others, are opposed to the use of GM crops in Africa. African producers have been receiving informal negative messages about adopting GM crop varieties from EC trade representatives over a substantial period of time (Oikeh, 2014). In the case of Burkina Faso withdrawing Bt cotton, described in Chapter 9, it now appears that the European anti-GM influence has gained wider influence.

The current debates over African agriculture's future discuss the possibility of decreasing dependency on the European stance on GM crops. Florence Wambugu, the regional director of ISAAA, reported that aid workers in Africa were 'brainwashed' about GM crops and told by their European offices not to promote the technology. She argued that while Europe is entitled to its opinion, it is 'dangerous' when Europeans tell Africans what to do (Wambugu, 2001). The International Food Policy Research Institute (IFPRI) report of 2013, for example, concluded that African regulatory models should be less influenced by European models, while some prominent figures in Africa, such as Dr Okogbenin from the African Agricultural Technology Foundation (AATF), argued that Africa should develop more independently from the European stance and start using biotechnology more widely (Okogbenin, 2014).

With regard to transgenic technology, while African farmers may appear to have less knowledge about GM technology than their European counterparts, they are very keen on getting seed varieties that can produce better yields, decrease costs and increase profits. They tend to have a pragmatic approach and might be willing to take a risk if the promise is good (Moon, 2014). They have recognised the potential benefits from using GM seeds but are worried about public opinion and potential liability issues resulting from the use of same on their farms (Richardson, 2002).

It also appears that in Africa, farmers are receiving mixed information about GM technology and, in a way, are 'getting tired of NGOs teaching them how to do agriculture' (Moon, 2014). This is how a farmer from Uganda described his experiences of dealing with NGOs:

> When we had just set our farm, about 30 years ago, we used to attend seminars organised by non-governmental organizations (NGOs), where we were usually taught about organic farming and the need for minimal use of agricultural chemicals as a way of sustaining the natural fertility of the soil and increasing crop production. We would also sometimes be warned about "fake" seeds – ensigo enkolelele – made by scientists and carrying the risk of causing cancers and depleting soil. We were even warned that these seeds were to be planted only once and that it would be useless to save seed from the harvested crop for replanting in the next season because the yields would be poor. The aim of the "fake seed producers", we were further told, was to keep us going back to the same people to buy seed every planting season. We had no understanding

at all of genetics and plant breeding . . . and we developed suspicions about
their efficacy and safety.

(Ssali, 2014, p. 36)

The author of the previous quote, Michael Ssali, is not an average farmer. He was
once a school teacher and works as journalist. He chairs the National Organization for Civic Education and Election Monitoring (NOCEM) and writes for the
Genetic Literacy Project.

Equally, African NGOs participate in the European debates. In 2013, Owen
Paterson, the UK's environment minister who is a GM supporter, claimed that GM
crops should be accepted in the UK, as they help to address hunger in developing
countries; he was opposed by NGOs from the anti-GM lobby bringing African
farmers' voices. In response to such claims, *The Guardian* newspaper published an
article by Million Belay, a representative of AFSA, quoting a farmer from Kenya:

It seems that farmers in America can only make a living from GM crops if
they have big farms, covering hundreds of hectares, and lots of machinery.
But we can feed hundreds of families off the same area of land using our own
seed and techniques, and many different crops. Our model is clearly more
efficient and productive. Mr Paterson is wrong to pretend that these GM
crops will help us at all.

(Belay, 2013, para. 10)

AFSA is a Pan African platform representing smallholder farmers, pastoralists,
hunter/gatherers, indigenous peoples, citizens and environmentalists from Africa
who possess a strong voice that shapes policy on the continent in the area of community rights, family farming, promotion of traditional knowledge and knowledge
systems, the environment and natural resource management (ABN, 2016).

Among its members are several African NGO networks and forums, such as the
African Biodiversity Network (ABN), the African Centre for Biodiversity (ACB),
COMPAS (Comparing and Supporting Endogenous Development), Coalition
for the Protection of Africa's Genetic Heritage (COPAGEN), The Eastern and
Southern Africa Small Scale Farmers' Forum (ESAFF), Fahamu, the Fellowship
of Christian Councils and Churches in West Africa (FECCIWA), the Indigenous
Peoples of Africa Coordinating Committee (IPACC) and African branches of two
anti-GM networks – Friends of the Earth and La Via Campesina (AFSA, 2016).

On the basis of the previous examples, it is possible to conclude that both
Europe and Africa influence one another in developing the debate on the use of
GM crops; the main object of such debates are African farmers.

Indian activism on GM crops in South Asia

Bhutan is an ancient royal kingdom in Tibet that is still a closed country. Due
to its geographic location and socio-political and religious status, the country's
most important neighbour providing political and financial support is India, while
China is a neighbour that is avoided (due to Tibetan tension).

For most outsiders, it has 'an almost mythical status as a real-life Shangri-La' where gross national happiness (GNH) measures the spiritual, physical, social and environmental health of its citizens and natural environment (Kelly, 2012). The original story of GNH goes back to the 1970s when the fourth king of Bhutan, His Majesty Jigme Singye Wangchuck, introduced the idea that Gross National Happiness mattered more than Gross National Product (GNH, 2016). It is also famous for another, more recent, political claim that it is a 100% organic country. This basically refers to the eradication of the use of chemical fertilisers and pesticides.

When in March 2015 I arrived to Navdanya organic farm in the Dehradun region for a course, I met a young Bhutanese man who was staying in Navdanya to help run the farm. He gave me a tour. It turned out that his father was an official in the Bhutanese government and a contact of Dr Vandana Shiva. While in Thimphu, the nation's capital, I was able to meet with Dr Saamdu Chetri in the office of the Gross National Happiness Centre. He is well educated and enjoys celebrity status outside of Bhutan. He worked previously in the private sector and in international development projects sponsored by European aid agencies, such as the Swiss Agency for Development and Cooperation and Helvetas, along with international organisations such as United Nations Development Programme (REALS, 2011).

Shiva's contribution to advocacy campaigns for organic agriculture began in 2010 with training farmers in the Samdrup Jongkhar province in south eastern Bhutan. The workshop was held by invitation of the Prime Minister of Bhutan, Lonchen Jimi Y Thinlay, in December of that year (Navdanya, 2016). The workshop was attended by 'hundreds of renowned persons of Bhutan' and led to the establishment of the initiative led by Dr Shiva to convert the province of Samdrup Jongkhar and, later, the whole country into adopting the organic way (Navdanya, 2016). The workshop was claimed as being part of a series of activities run by the Samdrup Jongkhar Incentive (SJI) which was put in place and led by a prominent Bhutanese lama, Dzongsar Khyentse Rinpoche, who is recognised as the third incarnation of the founder of Khyentse lineage of Tibetan Buddhism, in the Dewathang Chökyi Gyatso Institute in the Dewathang region. SJI was founded in March 2010 by Dzongsar Khyentse Rinpoche and calls itself a civil society organisation, or NGO. Its website claims that, alongside another educational project, 'SJI is a project of The Lhomon Society, [which is] a registered Civil Society Organisation (CSO) in Bhutan' (SJI, 2016).

In relation to the SJI workshop, the Bhutanese government has agreed to ban the use of agrochemicals in all 11 areas of the Samdrup Jongkhar Development Province and Dr Shiva was recognised to have led the process of banning the use of agrochemicals and transitioning to organic agriculture. Soon after the workshop, members of Navdanya arrived to Bhutan to train 200 farmers in organic agricultural practices. Selected farmers were then invited to Dehradun to stay at Navdanya farm in order to become ambassadors of organic knowledge and pass it on to their peers in Bhutan (Navdanya, 2016).

Through this process, Shiva has become the Bhutanese government adviser on organic agriculture. At the moment, there is no political debate in Bhutan regarding GM crops since the government has a clear-cut approach to minimising the risk of and exposure to GM crops in the country. Therefore, it might be possible to refer to NGO activism on GM crops, combined with the Indian influence (through Navdanya) on Bhutan, to explain why in 2011 the Bhutanese government passed the Cabinet Order banning the import, transit, release and research of GMOs, later passing the Biosafety Act in 2015 (Yangzom, 2015).

An important ally of Dr Shiva in Tibet is his Eminence, Prof. Samdhong Rinpoche, also referred to as Rinpoche. He served as prime minister of the Tibetan government in exile, having lived 48 years as a refugee from Tibet and working for His Holiness, the Dalai Lama. For a number of years, Rinpoche has contributed to the annual 'Gandhi and Globalisation' course held at Bija Vidyapeeth. The connection between Central Tibetan Administration (CTA) and Dr Shiva is quite established. She visited his Holiness in Dharsalama and consulted CTA on matters of organic agriculture. According to Rinpoche, all CTA settlement and settlement agricultural extension officers have been trained by Bija Vidyapeeth in organic farming who has called her a 'resource person' for CTA (Breasley, 2013). He explained his choice for Navdanya was due to its good reputation, its non-profit nature and affordability of its services and the practicality of organic agriculture (Breasley, 2013). As a result, CTA set a target to convert 'the entirety of . . . Tibetan settlements . . . into organic' (Breasley, 2013, para. 10).

Nepal is another country in the region which is influenced by the Indian activist. Dr Shiva has participated in the debate on GM crops there and Navdanya has collaborated with a number of Nepalese organisations. For example, Dr Shiva has even acted as founder of a gender unit in the (ICIMOD in Kathmandu (Hutanuwatr and Manivannan, 2005, p. 54).

Following the rhetoric on bio-piracy, Navdanya took part in the anti-Monsanto campaign in Nepal. This corresponded with both the national and regional interests of the organisation – a fight against agro-business, 'which is also called 'bio-piracy'. For example, Monsanto registered a patent on wheat with the European Patent Office which was developed from a traditional Indian variety found close to the border of Nepal and was being sold in the region. In 2004, the Research Foundation for Science, Technology and Ecology and Greenpeace filed a legal challenge against Monsanto and the patent was revoked (Shiva, 2006, pp. 148–149).

By 2011, Monsanto's Indian office had teamed up with USAID and the Gates Foundation in Kathmandu for a promotional launch of GM maize among Nepali farmers (Fuller, 2011). In turn, Nepalese civil society groups accused the government of 'foolishness' by inviting international players into the national government and began to organise protests. Farmers from the Kathmandu Valley organised a silent protest with banners against Monsanto in front of the US Embassy in Kathmandu, submitting a note to the embassy containing their demands. The protest included farmers associated with Navdanya (Navdanya, 2011). Following the earthquake in Nepal in April 2015, Navdanya launched a fundraising campaign to send '"seeds of hope" for the rejuvenation of food sovereignty and agriculture in Nepal' (Navdanya, 2015).

An NGO called Stop Monsanto in Nepal led the campaign outside the US Embassy. One member, Sabin Ninglekhu Limbu, claimed that the campaign was a success because it has, at least temporarily, delayed the promotion of GM seeds by USAID and resulted in the Nepalese parliament taking up the case with an investigation. He praised the success of the alliance of NGOs working in agriculture, particularly the role of Navdanya. While stressing that Stop Monsanto in Nepal is not a branch of Navdanya, he mentioned that one activist who had led the NGO during the campaign is an employee of Navdanya, so it was 'some kind of informal connection' (Limbu, 2012).

Some of the campaigns are only virtual, such as Monsanto Chahindaina which is a blog 'created to raise critical awareness about Monsanto's criminal involvement in agriculture and farming and organize against it' (Monsanto Chahindaina, 2011). It promotes a petition to 'prevent Monsanto from controlling the seed and controlling the people in Nepal'. It is also connected to another internet group called Stop Monsanto. Its main activity happens online through the sharing of information and is also supported by Vandana Shiva, who reposts its online entries.

However, not all anti-GM NGOs in Nepal have Navdanya links. International WWOOF (Worldwide Opportunities on Organic Farms) is 'a worldwide effort to link visitors with organic farmers, promote an educational exchange, and build a global community conscious of ecological farming practices'. WWOOF has a Nepalese branch which builds its vision on 'an intercultural dialogue on organic and sustainable agriculture farm' and runs visiting programmes and volunteer placements (WWOOF, 2016). WWOOF UK adopted the anti-GM stance, and in 2010, collaborated with such anti-GM NGOs as the Gaia Foundation, GM Freeze and celebrity chef Antonio Carluccio. At a joint public event, they promoted a blight-resistant potato developed by the Sarvari Research Trust from Wales. The event was featured in BBC show *The Urban Chef* (WWOOF, 2010).

While there is no obvious connection between WWOOF and Navdanya, volunteers interested in organic agriculture attend both Navdanya and WWOOF farms in Nepal and India. Both attract a similar kind of person and could possibly pass along similar ideas. The WWOOF coordinator published two blogs on Monsanto in 2013 (Regmi, 2013). In these, he referred to Basanta Ranabhat, a Nepalese activist working on anti-GM campaigns. Ranabhat is also mentioned in the chapter on Nepal in the book edited by Shiva (Shiva, 2016). Thus, the establishment of contact may be just one person away if it has not already been achieved. Interestingly though, in my e-mail correspondence with Fanindra Regmi on 26 February 2015, he referred to the WWOOF campaign against Monsanto as 'a nightmare' and avoided any further explanation (2015).

Advocacy science reciprocating at global level

The attempt of Monsanto to spread through Nepal is taken by some activists in the region as a threat to the entire Himalayan region (Shrivastav, 2015). While explaining the danger of GM crops, Arun Shrivastava referred to William Engdahl's book *Seeds of Destruction: The Hidden Agenda of Genetic Manipulation*, which portrayed GM seeds as part of the US strategy to control the world, and also wrote about a documentary by Bertram Verhaag. The book claims that USAID and Western foundations are used as a 'silent weapon' in that strategy. Scientists who have undertaken research and come to anti-GM conclusions are referred to as 'independent' and thus having experienced a 'well-structured and extraordinary coordinated attack' (Engdahl, 2007). William Engdahl is an American writer who previously 'worked as an economist and investigative freelance journalist in New York and Europe'. He is a Research Associate at Michel Chossudovsky's Centre for Research on Globalization in Montreal, Canada, and is a member of the editorial board of *Eurasia* magazine (Engdahl, 2016). Bertram Verhaag is a German documentary maker who produced a film entitled *Scientists under Attack – Genetic Engineering in the Magnetic Field*. Both Engdahl and Verhaag have presented GMO seeds as being part of the conspiracy outlined earlier where the US government and agro-chemical corporations attempt to control the world. Both refer to Árpád Pusztai, who is presented as a victim of such attacks on independent scientists. Engdahl has claimed that, according to Pusztai, it was Prime Minister Tony Blair who initially received an alarmed phone call from President Bill Clinton, who had approached Prof. James, who was the supervisor of Pusztai's project. Influenced by America, Blair made 'the promotion of GMO a cornerstone of his successful 1997 election campaign to "Re-brand Britain"', resulting in Prof. Pusztai's academic destiny being doomed (Engdahl, 2007, pp. 26–27).

Pusztai's case is a well-known early instance of advocacy science. Verhaag's film also featured the following: scientists Prof. Ignacio Chapela and Prof. Nina Fedoroff, journalist Jeffrey Smith who led the interviews, along with Dr Andrew Kimbrell. Kimbrell is an attorney and founder and director of the Center for Food Safety, who led the legal case against the FDA in terms of allowing GM food (Verhaag, 2010). All of the featured speakers in the film, with the exception of Prof. Fedoroff, have expressed concerns about the safety of GM food for human health and the motives of agro-chemical corporations in affecting the science behind the GM food industry (Verhaag, 2010). The first featured interviewee is Prof. Árpád Pusztai, who introduced his discovery about 'independent science' which has been suppressed by global agribusiness (Verhaag, 2010). The next scientist 'under attack' is Prof. Ignacio Chapela who had, prior to this film, already appeared in *The Future of Food* (a 2004 American documentary by Deborah Koons Garcia) and *The World According to Monsanto* (2008) by Marie-Monique Robin.

A Mexican-born microbiologist, Ignacio Chapela, is a professor at the Berkeley campus, University of California. Through his brother, he has established contacts with indigenous groups of Oaxacan Indians, Zapotecos and Chinatecos (Ross, 2004), which led him to serve as a scientific director for these groups. One theme within the initiative led by such grassroots units of farmers was GM crops. David Quist, a student of Prof. Chapela, was invited to conduct the project. Upon testing Mexican maize samples, they were found to be positive for the GM variety (GRAIN, 2003). In November 2001, *Nature* published a letter entitled 'Transgenic DNS introgressed into traditional maize landraces in Oaxaca, Mexico'. In this, they reported the presence of transgenic DN constructs in native maize landraces and expressed concern about 'the possibility of unintentionally transferring traits of ecological relevance onto landraces and wild relatives' (Quist and Chapela, 2001, p. 541). As in the case of Dr Pusztai, the article received severe criticism. Four of Chapela's colleagues submitted critical responses to *Nature*. One, Dr Metz, referred to the work of Quist and Chapela as 'mysticism masquerading as science' (Kaufman, 2002). Timothy Reeves, director of the International Maize and Wheat Improvement Center, argued that none of the testing conducted by the Mexican government discovered any genetically modified DNA in corn; however, he did agree that hypothetically such a problem of genetic contamination could appear (Kaufman, 2002). Prof. Chapela acknowledged that with this research he had 'stepped on other toes' within Berkeley University who were getting ready to host a generous grant from Novartis and the Mexican government. He reported death threats from a bureaucrat in the Mexican bio-security commission (Ross, 2004).

As with the Pusztai affair, the University of California tried to deny Chapela his tenure position but, unlike Pusztai, he fought back and was able to secure his professorship, which gave him the continuing freedom to research and promote his research findings and views on GM crops (Verhaag, 2010). As in the case of Pusztai, Chapela and his student attracted attention from anti-GM NGOs such as GMWatch and GRAIN, which published interviews with both scientists. In his interview, Chapela also referred to Greenpeace Mexico, which claimed that American farmers had 'dumped' GM corn in Mexico (Ross, 2004).

'Scientists under Attack' has briefly mentioned Dr Irina Ermakova and her research on rats and GM soya in Russia. She is a neuroscientist affiliated with the Russian Academy of Sciences at its Institute of Higher Nervous Activity. The issue of GM food came to her attention 'very accidentally' when she was asked to write an article for the wider audience about GMOs. She conducted a literature review and 'was struck by the imperfection of the research methodology used for creating GMOs'. It was 'clear' for her that GMOs created by such methods were capable of threatening living organisms (Ermakova, 2011). She then two published documents: the 'World Scientists' Statement: Calling for a Moratorium on GM (Genetically Modified) Foods and Crops', written in 1999 by 14 scientists including Dr Vandana Shiva and Prof. Árpád Pusztai; and the 'Open Letter from World Scientists to All Governments Concerning Genetically Modified Organisms (GMOs)' which was issued in 2000. The latter, signed by 815 scientists from 29 different countries who raised their concerns about GMOs, called for a ban on patents for life-forms and living processes that threaten food security. It further demanded the sanction of bio-piracy relating to indigenous knowledge and genetic resources and by such measures stop the violation of basic human rights for farmers and consumers (I-SIS, 2000). Among the signatories were Prof. Gilles-Éric Séralini and Dr Vandana Shiva.

After her literature review, Dr Ermakova wrote an article entitled 'Transgenetics – a new wave of evolution or a gene bomb?' and an open letter to the president of the Russian Federation (2011). She conducted her own laboratory tests involving the feeding of genetically modified soy to rats. In 2005, she presented her study at a conference organised by NAGS, a Russian NGO previously mentioned in

Chapter 4. Dr Ermakova, in her study, found that the mortality rates of rats fed with GM soya was six times higher than those given conventional soya. The study was mentioned in a leading Russian newspaper, *Pravda*, and the *Daily Mail* in the UK (GMO Compass, 2016). In December 2005, she presented at a conference entitled 'Epigenetics, Transgenic Plants & Risk Assessment' organised by Greenpeace and the Institute for Applied Ecology in Frankfurt am Main, Germany (Moch, 2006). In her paper, she referred to the fact that her findings were similar to those of Dr Pusztai and the research of Italian scientists under the leadership of Manuela Malatesta. The study found health hazards in mice fed with GM Roundup Ready soybean (Malatesta et al., 2002). That year, Dr Christopher Preston from the University of Adelaide criticised Ermakova's study in an article in *AgBioWorld*. His main criticism was around the lack of an explanation relating to the methodology used to conduct experiments, arguing that it was a case of 'only pseudo-replication' (Preston, 2005). In the style of Séralini, Ermakova created her own website – www.irina-ermakova.ru – on which she has published since June 2008 different entries on such topics as GMOs, UFOs and patriotic articles about modern Russia (Ermakova, 2016).

Klaus Ammann, an activist supporting GM crops, criticised Ermakova for presenting her un-reviewed results at a Greenpeace-sponsored conference and stated that her references are 'heavily biased' (Ammann, 2010, p. 4). He also updated his issue, commenting on her study by adding a letter he claimed she had written to Monsanto which 'starts with a positive declaration on GM crops, which contrasts heavily with her earlier condemnations' (Ammann, 2010, p. 4). Arguably, he has used the tactic of accusing his anti-GM opponent of 'sleeping with enemy'. In the letter released, it appears that Dr Ermakova approached Monsanto for funding to conduct further research. He provided two internet links to these letters. The link to her website is not valid, while the other leads to ask-force.org (Ammann, 2010, p. 4).

In 2012, Dr Ermakova published an article for a wider audience in the Russian journal *Federatsia* in which she explained the basics of GMOs and referred to both the Open Letter and World Scientists Statement and the studies of Pusztai, Malatesta, Quist and Chapela, Séralini. She included Russian scientists unknown to the English-speaking audience involved in the GM debate including Monastyrsky, Baranov, Yablokov, Kuznetsov, Kulikov, Olifirenko, Zydendambaev, Kopeikina and others. She repeated the argument of Engdahl, who she cited, that Pusztai was the first example of corporate business attacking independent scientists. She also mentioned that the research group led by Malatesta had lost a research grant due to its critique of GM crops (Ermakova, 2012).

The analysis of advocacy science, which its representatives call 'independent science', shows that it is a global phenomenon. Often, scientists who publish findings that refer to the negative impacts of GM crops are criticised in both their lack of academic robustness and social bias against GM crops. Most, with the rare exception of Dr Manuela Malatesta, who is by the way a co-author of G.-E. Séralini, become directly involved in activism, which is executed through establishing and strengthening links with the global network of NGOs working on GMO matters, and appearing in media. They also share similar reference sources and often have links to the same people and organisations (Dr Shiva and Greenpeace). This allows us to conclude that advocacy science regarding GM crops is a global pattern of transferring ideas across the global network. Scientist-activists acknowledge one another's findings and ideas which are exchanged and disseminated through their own networks.

Literature

ABN (2016) The Alliance for Food Sovereignty in Africa (AFSA). http://africanbiodiversity. org/alliance-for-food-sovereignty-in-africa-afsa/ as viewed 24.11.2016.

AFSA (2016) Members. http://afsafrica.org/category/about-us/members/ as viewed 24.11.2016.

Ammann, K. (2010) Are rat organs damaged after feeding on GM soybeans? AF-4 20090828. www.ask-force.org/web/AF-4-Ermakova/AF-4-Ermakova-20090828-web.pdf as viewed 22.11.2016.

BBC (2000) Greenpeace ambushes GM ship. *BBC Wales*, 25 February 2000.

Belay, M. (2013) GM crops won't help African farmers. *The Guardian*, 24 June 2013.

Breasley, A. (2013) Tibet goes organic. http://samdhongrinpoche.com/en/tibet-goes-organic-2/ as viewed 24.11.2016.

DeVries, J. (2013) Seed: Hope for smallholder farmers?, in B. Heap, D. Bennett (eds.) *Insights: Africa's future . . . Can biosciences contribute?* Cambridge: Lavenham Press.

Engdahl, F.W. (2007) *Seeds of destruction: The hidden agenda of genetic manipulation.* Quebec: Global Research. www.globalresearch.ca/seeds-of-destruction-the-diabolical-world-of-genetic-manipulation/25303

Engdahl, F.W. (2016) About F. William Engdahl: Biography. www.williamengdahl.com/about.php as viewed 22.11.2016.

EPO (1995) T 0356/93 (Plant cells) of 21.2.1995. www.epo.org/law-practice/case-law-appeals/recent/t930356ex1.html as viewed 20.11.2016.

Ermakova, I. (2011) Questions and answers. Online Entry 26.09.2011. www.irina-ermakova. ru/index.php/moi-publikatsii/515-387.html as viewed 22.11.2016.

Ermakova, I. (2012) GMO Test na Zrelost. *Federazia*, 1–3: 15–22.

Ermakova, I. (2016) www.irina-ermakova.ru

European Commission (2010) *Biotechnology report.* Brussels: DG for Communication.

Fuller, S. (2011) Why Monsanto? *Nepali Times*, Issue 577, 4–10 November 2011.

Gaskell, G. (2006) *Europeans and biotechnology in 2005: Patterns and trends.* Eurobarometer 64.3. Brussel: European Commission.

Gladwell, M. (2009) The tipping point: How little things can make a big difference. London: Abacus.

GMO Compass (2016) Study: GM Soy dangerous for newborns? www.gmo-compass.org/eng/news/stories/195.study_gm_soy_dangerous_newborns.html as viewed 22.11.2016.

GNH Centre Bhutan (2016) The story of GNH. www.gnhcentrebhutan.org/what-is-gnh/the-story-of-gnh/ as viewed 24.11.2016.

GRAIN (2003) With David Quist: The Mexican Maize scandal. Seedling, April 2003. www.grain.org/article/entries/367-with-david-quist-the-mexican-maize-scandal as viewed 22.11.2016.

Heijden, H.A. van der (2010) *Social movements, public spheres and the European politics of the environment: Green power Europe?* Basingstoke: Palgrave MacMillan.

Hoegel, J. (2014) Interview with Jens Hoegel. Brussels, 4 June 2014.

Hutanuwatr, P., Manivannan, R. (2005) *The Asian future: Dialogues for change*. Vol. 2. London: Zed Books.

ISAAA (2015) Pocket K No. 16: Biotech crop highlights in 2015. http://isaaa.org/resources/publications/pocketk/16/default.asp as viewed 22.11.2016.

I-SIS (2000) Pen letter from world scientists to all governments 1.09.2000. www.i-sis.org.uk/list.php as viewed 24.11.2016.

Jackson, L.A., Anderson, K. (2005) What's behind GM food trade disputes? *World Trade Review*, 4(2): 203–228.

Juncker, J.C. (2014) *A new start for Europe: My agenda for jobs, growth, fairness and democratic change*. Political Guidelines for the Next European Commission. Opening Statement. Brussels: EC.

Kaufman, M. (2002) Battlelines drawn in Mexico over genetically modified corn. *The Guardian*, 4 April 2002. www.theguardian.com/education/2002/apr/04/highereducation.internationaleducationnews

Kelly, A. (2012) Gross national happiness in Bhutan: The big idea from a tiny state that could change the world. *The Guardian*, 1 December 2012. www.theguardian.com/world/2012/dec/01/bhutan-wealth-happiness-counts

Limbu, S.N. (2012) Interview with Sabin Ninglekhu Limbu: Stop Monsanto in Nepal. *Global South*, 2(1): 44.

Lynch, D., Vogel, D. (2001) *The regulation of GMOs in Europe and the United States: A case-study of contemporary European regulatory politics*. New York: Council of Foreign Relations Press.

Maeckelbergh, M. (2009) *Learning from conflict: Arguing our way to a less hierarchical world*. Leiden: Institute of Cultural Anthropology and Development Sociology, Leiden University.

Malatesta, M., Caporaloni, C., Gavaudan, S., Rocchi, M.B., Serafini, S., Tiberi, C., Gazzanelli, G. (2002) Ultrastructural morphometrical and immunocytochemical analyses of hepatocyte nuclei from mice fed on genetically modified soybean. *Cell Structure Function*, 27(4): 173–180.

Moch, K. (ed.) (2006) Proceedings of the conference epigenetics, transgenic plants & risk assessment, December 1st 2005, Literaturhaus, Frankfurt am Main, Germany Freiburg. April 2006.

Monsanto Chahindaina (2011) https://monsantochahindaina.wordpress.com/

Moon, M. (2014) Interview with Marion Moon. Nairobi, 19 November 2014.

Moore, P. (2014) Interview with Patrick Moore. London, 30 January 2014.

Navdanya (2011) Navdanya joins movement in Nepal to stop Monsanto. www.navdanya.org/news/216-navdanya-joins-movement-in-nepal-to-stop-monsanto as viewed 24.11.2016.

Navdanya (2015) Seeds for hope for Nepal. www.navdanya.org/news/495-qseeds-for-hopeq-for-nepal as viewed 24.11.2016.

Navdanya (2016) Navdanya helps Bhutan to go organic. http://navdanya.org/news/143-navdanya-helps-bhutan-to-go-organic as viewed 24.11.2016.

Oikeh, S. (2014) Interview with Sylvester Oikeh. Cape Town, 10 November 2014.

Okogbenin, E. (2014) Contribution to the panel 'Food security, biotechnology & genetic modification: A mature conversation', Festival of Ideas. Cambridge, 16.05.2014.

Paarlberg, R. (2014) African non-adopters, in S.J. Smyth, P.W.B. Phillips, D. Castle (eds.) *Handbook on agriculture, biotechnology and development*. Cheltenham: Edward Elgar, pp. 166–175.

Pandor, G.N.M. (2014) Welcome speech. 13th International Symposium on the Biosafety of Genetically Modified Organisms (ISBGMO13), Cape Town. 9–13 November 2014.

Parr, D. (2014) Interview with Doug Parr. Oxford, 1 May 2014.

Preston, C. (2005) GM Soy affects posterity or misuse of science? *AgBioView*, 31 October 2005. www.agbios.com/main.php?action=ShowNewsItem&id=6980

Quist, D., Chapela, I. (2001) Transgenic DNA introgressed into traditional maize landraces in Oaxaca, Mexico. *Nature*, 414: 541–543.

REALS (2011) Dr. Saamdu Chetri. http://realsproject.org/team_member/dr-saamdu-chetri/ as viewed 24.11.2016.

Regmi, F. (2013) Monsanto, GMO seeds and the future of Nepalese farmers. https://wwoof-nepal.wordpress.com/2013/08/24/monsanto-gmo-seeds-and-the-future-of-nepalese-farmers/ as viewed 22.11.2016.

Richardson, D. (2002) A farmer's view, in B.J. Ford (ed.) GM *crops the scientists speak: Proceedings of the 2002 Cambridge Conference on genetically modified crops and food.* Cambridge: Rothay House, p. 12.

Ross, J. (2004) The sad saga of Ignacio Chapela, Anderson Valley Advertiser. www.theava.com/04/0218-chapela.html as viewed 24.11.2016.

Shiva, V. (2006) *Earth democracy justice, sustainability and peace.* London: Zed Books.

Shiva, V. (2016) *Seed sovereignty, food security: Women in the vanguard of the fight against GMOs and corporate agriculture.* Berkeley: North Atlantic Books.

Shrivastava, A. (2015) Weaponization of the food system: Genetically engineered maize threatens Nepal and the Himalayan Region. www.globalresearch.ca/weaponization-of-the-food-system-genetically-engineered-maize-threatens-nepal-and-the-himalayan-region/30512 as viewed 12.11.2016.

SJI (2016) Who we are. www.sji.bt/about/who-we-are/ as viewed 24.11.2016.

Ssali, M. (2014) Why I changed my mind about biotechnology for African countries?, in R.B. Heap, D.J. Bennett (eds.) *Viewpoints: Africa's future . . . Can biosciences contribute?* Lavenham: Lavenham Press, pp. 35–40.

Van Montagu, M. (2013) The irrational fear of GM food. *The Wall Street Journal*, 22 October 2013. www.wsj.com/articles/the-irrational-fear-of-gm-food-1382481840

Van Montagu, M. (2015) Interview with Marc Van Montagu. *Trends in Plant Science*, 20(6): 325–327.

van Schendelen, R. (2005) *Machiavelli in Brussels: The art of lobbying the EU.* Amsterdam: Amsterdam University Press.

Verhaag, B. (2010) Scientists under attack: Genetic engineering in the magnetic field of money. www.youtube.com/watch?v=ADNE1B2RI5Y

VIB (2013) International Plant Biotechnology Outreach (IPBO). www.vib.be/en/research/scientists/Pages/IPBO.aspx as viewed 22.11.2016.

Wambugu, F. (2001) *Feeding Africa. Geneflow: A publication about the Earth's plant genetic resources 2000–2001.* Fiumuchino: IPGRI.

Weasel, L. (2009) *Food fray: Inside the controversy over genetically modified food.* New York: AMACOM.

WWOOF (2010) *WWOOF UK News*, Issue 227, Summer 2010. www.wwoof.org.uk/sites/wwoof.org.uk/files/WWOOF%20Issue%20227.pdf as viewed 22.11.2016.

WWOOF (2016) Welcome to WWOOF Nepal. https://wwoofnepal.net/

Yangzom, T. (2015) Update on biosafety regulation Bhutan. A paper presented at the 3rd South Asia Biosafety Conference, Dhaka. 19 September 2015.

11 Conclusions

We live in the times when food security is a part of global challenges, including climate change, depletion of natural resources and growing political instability. Genetically modified crops are a modern phenomenon that exists in today's world alongside any other inventions created with the progress. Yet it is fascinating that GM crops have caused such a strong social debate with a heated polemic. The fabric of political ideas associated with the debates on GM crops is very vast.

In order to better understand and classify these ideas, in this book I distinguished six broad discourses which cover major ideas on GM crops: nature fundamentalism, post-colonialism, regionalism, sustainability, alterglobalism and feminism. They offer different arguments in favour or against GM crops. While two – post-colonialism and sustainability – discourses allow to frame arguments for both sides – proponents and opponents of GM crops – the remaining four discourses are used by GM opponents only. The major themes which appear throughout all discourses are discussion of future models of agriculture with particular focus to smallhold farmers and impact on environment, equality (political, economic, gender) and public trust in science. Most discourses are dimensional – they are framed as alternative, a critique to the mainstream approach (conventional agriculture, mainstream genetic science, mainstream aid policy and a traditional gender hierarchy model). A historic approach has allowed us to dig deeper and identify the sources for ideological inspiration. What is striking though is that certain ideas have been modified over time. For example, the proposition that plants can feel was not offered by Rudolf Steiner, just as Rachel Carson did not criticise Bt crops. These statements have later evolved in the ideology of their followers. Table 11.1 summarises all six discourses and the major ideas, representatives, campaigns and attitudes towards GM crops they represent. One can see that there are overlaps: activists and organisations they represent cover in some cases more than one discourse. It is also safe to say that Dr Vandana Shiva is the most cited activist and has the most intense networks with NGOs working on the subject worldwide. Her social activism and strong presence in the debate on GM crops led her to develop enough credentials to influence governments in South Asia, such as Nepal, Bhutan and the government of Tibet in exile. However, in her own country, India, she has much lesser political influence. It is also the case that recently in 2015 Indian government banned the Greenpeace Indian branch from receiving foreign donations (Sherwell, 2015).

Table 11.1 Summary of analysed discourses

Name of the discourse	Major themes	Participating organisations	Associated partners	Sources of ideas	Representatives	Campaigns	Attitude to GM crops
Nature Fundamentalism	Organic 'biodynamic' agriculture, alternative science versus reductionism	IFOAM, EKAH, SAG	GMO-Free Europe, GENET, GMWatch, GRAIN, PAN, Third World Network	R. Steiner's ideas, anthroposophy, shamanism	Florianne Koechlin	Beobachter Initiative, the Gen-Schutz Initiative	Negative
Post-colonialism	Independence of former colonies, 'indigenous' knowledge national development, social utopia	(1) The Research Foundation for Science, Technology and Ecology, Navdanya	Brod für die Walt, ActionAid, AFSA, AGRA Watch Schumacher College, UK	Mohandas Gandhi, Sir Albert Howard	Vandana Shiva	The GM banana campaigns	Negative
		(2) The African Centre for Technology Studies	The Bill and Melinda Gates Foundation, AGRA	Julius Nyerere	Calestous Juma		Positive
Regionalism	Food sovereignty, national agriculture, protectionism, food safety	Danube Soya Association, NAGS	GENET, GMO-Free Europe, Hraniteli, Pravda o ede	La Via Campesina	Matthias Krön, Elena Sharoykina	Public events and digital campaigns. 'No Monsanto. Russia without GMOs.'	Negative

Name of the discourse	Major themes	Participating organisations	Associated partners	Sources of ideas	Representatives	Campaigns	Attitude to GM crops
Sustainability	Three pillars of sustainable development (environment, economy and social development) Cultural diversity as fourth pillar	European Federation of Biotechnology, Earth Open Source	AgBioWorld Foundation LobbyWatch	Brundtland Report	Klaus Ammann	Declaration of the scientists in Support of Agricultural Biotechnology	Positive
Alterglobalism	'Agrarian citizenship', de-globalisation, alternative development	Focus at the Global South, Navdanya, Confédération Paysanne	La Via Campesina, Friends of the Earth, Slow Food movement	The Seattle battle	Vandana Shiva, Walden Bello, Jose Bove	Campaigns against Monsanto	Negative
Feminism	Female emancipation, environmental conservation	Navdanya, NAGS	FINRRAGE, WECF, ICIMOD, WEDO	Francoise D'Eaubonne, James Lovelock	Vandana Shiva, Elena Sharoykina	Diverse Women for Diversity	Negative

Reflections on the discourses

The discourse over the confrontation of science and **nature fundamentalism** proves that modern science has diverse values, both personal and moral. There seems to be a confrontation between the two discourses – science fundamentalism against nature fundamentalism, as the first promotes superiority of scientific knowledge and the other connects to the desires shared by some scientists to be able to embrace spiritual, unprovable or 'unscientific' knowledge and add emotions and moral values to research. Francis Crick, one of the discoverers of DNA, once said:

> The view of ourselves as [with no soul] 'persons' is just as erroneous as the view that the sun goes around the earth, this sort of language will disappear in a few hundred years. In the fullness of time, educated people will believe there is no soul independent of the body, and hence no life after death.
>
> (Silver, 2006, p. 16)

However, Crick was wrong in suggesting that every scientist wishes to have the same belief. For example, scientists and activists including Dr Koechlin at the Berlin Conference have referred to Crick's 'central dogma of molecular biology' and discussed how their research came as 'unwanted information'. Despite their education, there are scientists who do not want to accept a worldview which is based purely on mainstream science. They require a spiritual component and mystery which is not uncovered by science. Dr Koechlin, who studied biology and chemistry and understands the mechanics of gene transfer, believes in the complexity of life that cannot be embraced by modern science. This comes as a personal choice. And she is not the only one. American biologist Prof. Leon Kass is another example of the scientist who finds modern science being 'overconfident of its ability' to explain life and who wants to believe that 'life and soul are irreducibly mysterious' (Kass, 2002, p. 292).

The organic agriculture movement, with its ideological roots in Steiner's contradictory combination of mysticism and science, seems to be a particularly good fit for those unsatisfied with mainstream science, scientists and members of the wider community who do not accept scientific views on life and instead focus on biogenetics. The urge for mystery and the inclusion of emotions is a good match within the non-governmental format. NGOs are good at communicating with people; they choose the language the general public understand and wish to speak, reaching their audience through formal and informal networks, allowing new members to become involved and speak about their concerns and fears. It can be argued that it is not a coincidence that anthroposophy and biodynamic agriculture ideology feed into the anti-GMO movement and that their organisations provide structures to quickly mobilise resources to run campaigns against GM crops.

The **post-colonial discourse** has emerged through the search for national post-colonial strategies in liberated former colonies of global metropolis in Asia and

Africa. Yet the presence of former metropolis in the current debate is clearly identifiable. In fact, NGOs and IOs from developed countries tend to lead the debate about future development of agriculture in the developing world. While the US has a pro-GM stand, European politicians and NGOs tend to send mixed messages and this complicates the debate. Another important tool of influencing the debate in Asia and Africa is funding; it is used for scientific development of new GM varieties designed to meet the challenges in tropical climates and also finances to cover campaigns, visits of activists and publications. While two compared activists Vandana Shiva and Calestous Juma have different assessments of GM crops, they share one important thing in common – they wish for more technological and political independence for their countries. How much of this desired independence is possible under the global regime of transfer of ideas through formal institutions (international aid) and informal (NGOs) remain under question. Both activists draw the financial support from the West: Juma, from the US, and Shiva, from Europe; hence, it may also contribute to explaining why they have different assessments of GM crops. Equally, they use social utopias offered by Gandhi and Nyerere as the foundation for their ideas. These visionary ideas are utopian, but at least they are 'indigenous', home bred and not imposed from outside. From that perspective, the debates on GM crops also tackle a big issue – the question of foreign aid. Is it used as a foreign policy tool by wealthy nations and, if so, what are their motives?

Greenpeace leader Kumi Naidoo, who also claims to be born and raised in Africa, considers international development strategies to use GM crops to solve global hunger as 'PR gloss for a corporate agenda' and an intentional attempt to destroy traditional farming, which would basically mean political and economic control over developing world (Naidoo, 2013). Thus, for him the humanitarian aspirations to feed the world with GM crops is a façade for international power games.

Another discourse which continues the discussion of international relations about power asymmetry is **regionalism**. In this discourse, European countries and regions focus on ensuring their food security which is taken not as availability of food, but as food sovereignty, meaning the availability of food for nationals of a country or a region should not imply vulnerability in relating to international partners. This explains a relatively suddenly sharp redirection of Russian food security strategy as anti-GM. Of course, one can argue that the publication of the Séralini study results has affected the position of the Russian government from the point of food safety of GM crops. But it is also most possible that current tensions with the West and sanctions are the major reason for this policy change. The anti-GM rhetoric placed in the regional discourse also gives a rare place for manoeuvring and protecting regional farmers under the global liberal trade regime that generally does not allow such support. Arguably, the case of Danube Soya is a case of a politically accepted protectionist policy, as it is based on the grounds of food safety and a globally growing demand in soya rather than creating unfair competition. Yet creating the voluntary labelling scheme for non-GM Europegrown soya opens new opportunities for German-speaking business and farmers. Let me call it a smart protectionism of the twenty-first century.

Sustainability is an interesting discourse, as it covers all arguments of GM crops: food safety, environmental impact of the new crops, political and social justice and economic profitability of such crops.

Although both sides operate in the same political ideological framework of sustainable development, which is not new and has witnessed so many attempts to be defined and redefined, they cannot reach compromise about the GM crops. In many cases, the same argument can be used for and against to the same end. It appears that despite the common objective of sustainability, the debate on GM crops has only been deconstructed in smaller debates based around the three topic areas of sustainable development, environment, economics and social development, with much attention given to the two sub-issues of sustainable development, biodiversity loss and food security. As a result, the opposing sides have not been able to reach any agreement.

The concept of sustainable development itself is very broad and allows many interpretations which at times may be confusing and has been substantially criticised (Adams, 2001). It is both a concept and a discourse and, arguably, the most salient discourse in the current global development framework and policymaking. It is ubiquitous and has been influencing institutional capacity building and policymaking for at least the last three decades. So the main problem with this discourse though is its very broad scope and assigning of values which allow different sides to prioritise different issues and values within this discourse. Such approach also allows the argument that most activists do not wish to arrive to a compromise and sign 'peace' with the opponents. That's possibly why it becomes such an important event when a rare case of activists changing sides happens. It validates 'the rightness' of the side and confirms 'the wrongness' of the opponents.

Alterglobalism is a discourse which directly tackles globalisation – arguably the source of many challenges discussed in the discourses introduced earlier – post-colonial development, regionalism and sustainability. Alterglobalism is similar to post-colonialism as it fights against cultural imperialism. Both discourses have named corporations like Monsanto their enemy. But it is a much more diverse discourse, as it covers the global justice movement and regional, mostly European, fights to protect cultural diversity, through such tools as culinary capital. It is a unique discourse, as it produced the movement of movements, and it is a very powerful 'glue' that created a global network which allows the passing on and dissemination and co-creation of new ideas.

Feminist discourse might appear to be on a margin. Yet it is also a discourse which is used in the debate to argue against the use of GM crops. The issue of gender inequality is taken as a male domination over Nature. Thus, the true justice was to allow women conducting agriculture in a traditional manner which would allow the saving of natural resources and fair distribution of food in the world.

Finally, let's imagine what one can expect in the future for the debates on GM crops. Over the years, activists from both sides around the world have taken part in the public debates against each other in the best tradition of gladiators from ancient Rome. However, the opponents do not have any consensus reached so far. Furthermore, one can expect only another continuation of the debate. At some point, it looks like the anti-GM side is winning, particularly in Europe, and they are ready to reinvigorate and protect the position at any moment (It is my personal observation conducted at The Colour of Science Conference, Berlin, 2014). At the same time, European position is complex. The European Commission sponsors the research on GM crops, and there are influential political lobbies such as EuropaBio to advocate for biotechnology. At the EU level, countries also vary in their approach. With new political processes towards de-integration, such as Brexit, there will be a search for new, alternative developmental strategies for national agriculture. Some even argue that Brexit will turn the UK into a GM country, and 'as part of the preparations for EU exit' the government was looking at 'possible future arrangements for the regulation of genetically modified organisms' (Wilkinson, 2016, para. 4).

Globally, things are also complicated. On the one hand, there are countries moving towards commercialisation of GM crops. According to ISAAA, most GM crops are grown now in the developing world. On the other, Burkina Faso intends to give up its GM cotton. The reasons are different, but the source of political lobby is the same. It is done through NGOs.

Through this review of six discourses, I hope to persuade the reader how the debates on GM crops are centrally located in the discussion of the modern and future political, economic and social development of our world. The stakes are high indeed. Food is the basis for any policy. No food, no political stability. And there are no clear answers. The answers should be given by science, but if there are different sciences (corporate science, civil society science, advocacy science), whom can we trust? And should we trust at all? But if we don't trust, how can we know what is good and what is bad, what is working and not working? I personally find it difficult not to trust science. I do want to trust science.

Another important feature of the debate, which has served the main argument of the book, is the role of non-governmental organisations. While some discourses, such as regionalism, show that there is a space for state as political actor, it also seems the case that NGOs act as the major channel for lobbying in the GM debate at the global level. After all, only in this research I identified 358 NGOs associated with the debate. Of course, there are more out there.

The review of the discourses and different tactics of conducting debates (changing sides, creating its own science and international transnational networks) show that these organisations have a capacity for lobbying both in favour of and against GM crops. Most of the activists involved may or may not have scientific education, but they will claim these credentials at some point. It is also true that, for many, this is their professional activity, a job, a business and often an ideology they worship. To understand where is the boundary between a job, a political interest and personal values and ideas is almost impossible.

What is next?

What, we, the observers, can take from the debate is something very obvious. It is difficult to take a political decision on GM crops. Often I am asked whether it is safe to eat GM, and as a social scientist, I do not have an answer. The same replies to policymakers and the general public.

The use of emotional language and the confrontational, aggressive form of debates may entertain those of the public who watch them, but it is not productive in terms of negotiating pragmatic, working solutions for the challenges of agriculture. At some point, I considered that, despite the continuing excessive debates on GM crops, there is still the need for a more balanced, less biased and emotionally cool debate that assesses the arguments both for and against GM crops.

Here is one:

> GM crops elicit responses ranging from claims of miracle products which alleviate poverty at a single stroke at one end of the spectrum through to the opinion that they will devastate agriculture and the environment. The reality lies between these extremes. GM crops are not miracle products for poverty alleviation, but neither is there the evidence that they will cause the scale of damage associated with indiscriminate use of pesticides and fertilizers or indeed with the removal of hedgerows and woodlands to produce larger field sizes.
>
> (Morse and Mannion, 2009)

Another example is the Montpellier Panel Report, produced by a group chaired by Prof. Gordon Conway, which has argued for sustainable intensification in agriculture. Such intensification aims to produce greater nutritional yields with less pesticides, fertilisers and emissions together with better use of environmental resources. Ecological intensification, such as intercropping and the push-pull system of pest control, conservation farming and genetic intensification which includes conventional breeding, cell and tissue culture, marker-assisted selection and genetic modification, accompanied by socio-economic intensification, are proposed as three possible solutions to farmers' challenges (The Montpellier Panel, 2013). As one can see, the GM technique is not praised by the report as a 'silver bullet' for food security nor are its results denied, but it is proposed as one of the many possible strategies to be considered.

To sum up, I would argue that the main function of the GM debate, particularly as they are conducted by NGOs, is not to arrive at a compromise or a balanced solution about how to handle GM plants. This is a modern day democracy, if I may say so. The debates represent an opportunity for different groups to voice their views and use it in lobbying the political interests they support. They vary widely in their opinions, values and discourses. It is a rainbow with different colours and attitudes, and it will be impossible to bring them to one conclusion.

Literature

Adams, W.M. (2001) *Green development: Environment and sustainability in the Third World.* London: Routledge.

Kass, L. (2002) *Life liberty and defence of dignity: The challenge of bioethics encounter.* San Francisco: Encounter Books.

Morse, S., Mannion, A.M. (2009) Can genetically-modified cotton contribute to sustainable development in Africa? *Progress in Development Studies,* 9: 225–247.

Naidoo, K. (2013) GM crops: The genetic colonialists. *The Guardian*, 24 June 2013.

Sherwell, P. (2015) India bans foreign funds for Greenpeace in latest crackdown on charities with Western ties. *The Telegraph*, 4 September 2015.

Silver, L.M. (2006) *Challenging nature harder*. New York: Collins.

The Montpellier Panel (2013). *Sustainable intensification. A new paradigm for African agriculture*. London: Imperial College.

Wilkinson, M. (2016) British farmers could grow GM crops after Brexit, reveals minister. *The Telegraph*, 26 October 2016.

Appendix 1
List of NGOs participating in the GM debate

Business associations

AfricaBio
AgBioWorld Foundation
Agriculture and Horticulture Development Board (AHDB)
British Society of Plant Breeders Ltd
CropLife International
Danube Soya
EuropaBio
European Biotech Forum
European Federation of Biotechnology
International Service for the Acquisition of Agri-biotech Applications (ISAAA)
Irish BioIndustry Association
National Non-food Crops Centre Board
National Union of Agropastoral Workers (Syntapa)
Supply Chain Initiative on Modified Agricultural Crops (SCIMAC)
Washington Biotechnology Action Council
World Business Council for Sustainable Development (WBCSD)

Farmers' associations

Agricultural Renewal in India for Sustainable Environment (ARISE)
Albanian Association of Organic Horticulture-Bioplant
Albania Permaculture Resource Centre
Andhra Pradesh Farmers Association
Appiko
Association des Producteurs de Coton Africains (AProCA)
Beej Bachao Andolan
Biodynamiska Föreningen
California Certified Organic Farmers
Centre for Indian Knowledge Systems (CIKS)
Centre National des Jeunes Agriculteurs (CNJA)

Coldiretti Italian Farmers' Union
Confédération Nationale des Syndicats de Travailleurs Paysans (CNSTP)
Confédération Paysanne (CP)
Deshma Sikshan Sansthan
East Africa Farmers Federation (EAFF)
Eastern and Southern Africa Small Scale Farmers' Forum (ESAFF)
Ekologiska Lantbrukarna (Swedish Association of Ecological Farmers)
European Farmers and Agri-Cooperatives
Fédération Nationale des Syndicats d'Exploitants Agricole (FNSEA)
Federation of Farms in the Republic of Macedonia
Food and Allied Workers Union
German Farmers' Association
Green Foundation
Indian National Trust for Art and Cultural Heritage (INTACH)
Irish Organic Farmers and Growers Association
Irish Cattle and Sheep Farmers' Association
Janhit Foundation
Jharkhand Alternative Development Forum
Karnataka Farmers Movement (Karnataka Rajya Raitha Sangha)
Kheti Virasat
La Via Campesina
Minnesota Food Association
Mother Earth Farm
National Federation of Ecological Agriculture (FNAE)
Navadarshanam
Network of Farmers' and Agricultural Producers' Organisations of West
   Africa (ROPPA)
Norges Bondelag
Norsk Bonde-og Småbrukarlag
Parisaran Samvrakshana Kendra
Plateforme sous Régionale des Organisations Paysannes d'Afrique Centrale
   (PROPAC)
Prakruti Paramparik Bihana Sangharakshana Abhijan
Sirmaur Farmers Association
Tamil Nadu Organic Farmers Association
Vidarbha Organic Farmers Association
Vidarbha Jan Andolan Samiti
Vrihi Bija Binimoy Kendra
The Women's Alliance

Foundations

African Agricultural Technology Foundation (AATF)
AME Foundation
Arche Noah

Avalon Foundation
Bill and Melinda Gates Foundation
Biosustain DTU Novo Nordisk Foundation
Consultative Group for International Agricultural Research Fund
David Suzuki Foundation
Deutsche Gesellschaft für Technische Zusammenarbeit (GTZ)
Ford Foundation
Foundation for Environment and Agriculture (FEA)
Foundation on Future Farming
Foundation for the Philippine Environment
GAIA Foundation
Glaser Progress Foundation
Goa Foundation
Green Foundation
Hivos International
Hungarian Environmental Partnership Foundation
Indo-German Bilateral Project
ProTerra Foundation
The Rockefeller Foundation
Small Planet Fund (formerly the Rudolf Steiner Foundation)

Humanitarian NGOs

ActionAid
Caritas Internationalis
Catholic Agency for Overseas Development (CAFOD) (London, England, UK)
International Federation of Red Cross and Red Crescent (IFRC)
Oxfam
Save the Children

International think tanks

Association of Environmental and Resource Economists
Center for Alternative Development Initiatives
Centre for Food Safety
Centre for Science in the Public Interest
Chatham House
Council for Responsible Genetics
EcoNexus
Environmental Policy Institute
Foundation on Economic Trends
Genetic Rights Foundation
Green Alliance
Institute for Agriculture and Trade Policy (IATP)

Institute for Food and Development Policy/FoodFirst
Institute of Ideas
Institute of Plant Biotechnology Outreach
International Forum on Globalization
International Institute for Sustainable Development
New Economics Foundation
People-Centered Development Forum

Green/social and environmental justice NGOs

Action for Solidarity Equality Environment and Diversity (ASEED Europe)
African Biodiversity Conservation and Innovations Centre
African Biodiversity Network (ABN)
African Centre for Biosafety
Agent Green
AGRA Watch
Agrolink
Alliance for Bio-integrity
Alliance for Food Sovereignty in Africa (AFSA)
Allow Golden Rice Now
Anthrosana
Ärztinnen und Ärzte für Umweltschutz
Association for India's Development
Association pour la Taxation des Transactions pour l'Aide aux Citoyens
   (ATTAC)
Basel Appeal against Genetic Engineering
Bio Alpe Adria
Bioengineering Action Network
Biolands
BioRe India
Biorespect
Biosustain
BioWatch
Bird Watch
Bread for the World
Bundesverband Naturkost Naturwaren (BNN)
Burgers voor gentechvrij voedsel
Campagna per un Comune Antitransgenico
Centre for Ethics and Toxics
Centro Internazionale Crocevia
Christian Ecology Link
Clean Water Fund
Coalition for the Protection of African Genetic Heritage (COPAGEN)
Coalition of Free GM India

Coalizione ItaliaEuropa – LIBERI DA OGM
Collectif d'Action GenEthique (CAGE)
Community Alliance for Global Justice
Comparing and Supporting Endogenous Development (COMPAS)
Compassion in World Farming
Concerned Citizens
Confederación Española de Consumidores y Usuarios
Connecticut Coalition for Environmental Justice
Consiglio dei Diritti Genetici
Consumers International
Coordinadora de Organizaciones de Agricultores y Ganaderos (COAG)
Corner House
Corporate Europe Observatory
Corporate Watch
CRANN (Centre for Research on Adaptive Nanostructures and Nanodevices)
Cretan Network Against GMO
CRIIGEN (Comité de Recherche et d'Information Indépendantes sur le Génie Genétique – Committee for Research and Independent Information on Genetic Engineering)
Council for Responsible Genetics
Czech Technology Platform for Organic Agriculture
De Gentechvrije Burgers
Dio
Diverse Women for Diversity (DWD)
Earth First!
Earthlife Africa
Earth Open Source
Ecologistas en Acción
Ecoropa
Eco-sense
EcoSolidar
Emweltberodung Lëtzebuerg (EBL)
Ending Destructive Genetic Engineering (EDGE)
Environment Support Group
Equivita
ETC Groups (Action Group on Erosion, Technology and Concentration)
European Network of Scientists for Social and Environmental Responsibility (ENSSER)
Fahamu Trust
Family Farm Defenders
Fellowship of Christian Councils and Churches in West Africa (FECCIWA)
Feminist International Network of Resistance to Reproductive and Genetic Engineering (fiNRRAGE)

Focus at Global South
Food and Trees for Africa
Food First
Friends of the Earth
Gateway Green Alliance
Gene Campaign
Gen-ethisches Netzwerk
Genetic Engineering Action Network
Genetic Forum
Genetic Resources Action International (GRAIN)
GenetiX Snowball
GeneWatch
Gentechnikfrei Europa
Global Commons Institute
GMO Free Idaho
GMO-Fritt Norrbotten
GMO-NET
Greek Network against GMOs
Green Network of Vojvodina
Greenpeace
Group for Environmental Monitoring
Hraniteli
IG Saatgut (Initiative for GE-free Seeds and Breeding)
Indigenous Peoples Coalition against Biopiracy
Indigenous Peoples of Africa Co-ordinating Committee (IPACC)
Initiative Magnesia
Institute for Agriculture and Trade Policy (IATP)
Institute for Social Ecology
Interessengemeinschaft für gentechnikfreie Saatgutarbeit (IG Saatgut)
International Center for Technology Assessment
International Organization of Consumers Union
International Union for Conservation of Nature (IUCN)
Irish Seeds Savers Association
Just Label It
Kansalaisten Bioturvayhdistys
Kheti Virasat
Kleinbauern-Vereinigung
Legambiente
Les Amis de la Terre
Lhomon Society
Linking Environment and Farming (LEAF)
London Food Commission
Luleå Biodlarförening
Matupproret

Mellifera
Myrada
National Association for Genetic Safety
National Environmental Trust
National Family Farm Coalition
National Organization for Civic Education and Election Monitoring
National Sharecroppers Union/Rural Advancement Fund International (RAFI)
National Wildlife Federation
Native Forest Network
Natura Balcanica
Nature et Progrès Belgique
Navdanya International
Nederlands Platform Genetechnologie
Network for Safe and Secure Food Environment
No Patents on Life
Norfolk Genetic Information Network
Norges Bygdekvinnelag
Norges Bygdeungdomslag
Norges Miljøvernforbund
Norsk Landbrukssamvirke
No Thanks
OGM Danger
Oikos
Organic Agriculture Association in Africa
Organic Consumers Association (OCA)
Participatory Ecological Land Use Management Association (PELUM)
Pesticide Action Network (PAN)
Pesticide Action Network North America (PANNA)
Planetum
Plataforma Galega Antitransxénicos
Plataforma Transgénicos Fora
Primal Seeds
Priroda I Molodezh
PRObio
Pro Natura
ProSpecieRara
Public Citizen's Global Trade Watch
Rashtriya Yuva Dal
Resistance Is Fertile
Right2Know
Rising Tide UK
RSPB (The Royal Society for the Protection of Birds)
SAG (Schweiz Arbeitsgruppe Gentechnologie: Dachorganisation aller gen-
    techkritischen schweiz)

Samdrup Jongkhar Incentive (SJI)
Save Our Seeds
Schweizer Tierschutz STS
SeedSave
Shetkari Sanghatana
Sierra Club
Sluzba Okhrany prirody
Som Lo Que Sembrem
Sotenäs Biodlareförening
South African Committee on Genetic Experimentation
South African Council of Churches
South African Environmental Justice Networking Forum
South African Freeze Alliance on Genetic Engineering
South Durban Community Environmental Alliance
Southern Action on Genetic Engineering (SAGE)
Southern Eastern Environmental Network
Spanish Farming Family Association
StopOGM-Alliance suisse pour une agriculture sans génie génétique
STOP Monsanto in Nepal
Tanzania Alliance for Biodiversity
Third World Network
Totnes Genetics Group
Umanotera
Uniterre
Upper Midwest Resistance against Genetic Engineering (RAGE groups)
Uralskyi Ekologicheskiy Souz
Utviklingsfondet
Utviklingsfondets ungdomsorganisasjon
Velt
Verband für organisch-biologischen Landbau Bioland
Verband Lebensmittel Ohne Gentechnik – VLOG
Verdi ambientee societa (VAS)
Vermont Genetic Engineering Action Network
Vo imya Zisni
Vosrozdenie Zoloti Vek
Vsa Pravda o ede
Wervel
Western Organic Network
Women in Europe for a Common Future (WECF)
Women's Environmental Network
Women's Environment and Development Organization (WEDO)
World Council of Churches
World Wide Opportunities on Organic Farms (WWOOF)
World Wildlife Fund

Za biobezopasnost
Založba AJDA
Zelena Akcija
Zelenaya Orbita
Zelenite
Zukunft säen!, Montézillon
Zukunftsstiftung Landwirtschaft
Zürcher Tierschutz, Zürich

Scientific research and science lobbying organisations

African Centre for Technology Studies
AgBioWorld
Alliance for a Green Revolution in Africa (AGRA)
American Association for the Advancement of Science (AAAS)
Biosustain DTU Novo Nordisk Foundation
Bioversity International
Cornell Alliance for Science
Earth Open Source
Genetic Literacy Project
GMO Seralini
Institute of Arable Crops Research/Rothamsted Centre
Institute of Plant Biotechnology Outreach (IPBO)
Institute of Science in Society
International Centre for Integrated Mountain Development (ICIMOD)
International Crops Research Institute for the Semi-Arid Tropics (ICRISAT)
International Food Policy Research Institute
International Institute of Tropical Agriculture (IITA)
International Maize and Wheat Improvement Center
International Rice Research Institute
International Service for the Acquisition of Agri-biotech Applications (ISAAA)
John Innes Centre
Kenya Agricultural Research Institute (KARI)
Mae-Wan Ho Institute for Science in Society (ISIS)
National Agricultural Research Organisation (NARO)
Public Eye on Science
Sarvari Research Trust
Schumacher College
Science Literacy Project
Sense about Science
Society for the Anthropology of Food and Nutrition (SAFN)
Testbiotech
Union of Concerned Scientists

Social movements

Demeter International
Ecology and Feminism Movement
March Against Monsanto
Occupy Movement
Organics International (IFOAM)
Slow Food International
Tanzania Organic Agriculture Movement

Appendix 2

TODAY IS INTERNATIONAL MARCH AGAINST MONSANTO, there are millions of people marching today in most of the main cities across the globe. WHY IS THERE NO MEDIA COVERAGE?

Be the change you wish to see in the world. Please show this piece of paper to at least 1 other person. Knowledge is power.

Monsanto are the largest seed company in the world. This means that they own our food chain, and farm animal's food chains. This makes them bigger than guns and bigger than war. Most people have not even herd of them.

Monsanto used to be a chemical warfare company, they are most famous for their creation Agent Orange. They now make "pesticides" such as ROUND UP.

ROUND UP and their other pesticides are very toxic and kill all plants, animals and insects that are on the land it is sprayed on. It does not break down easily thus rendering the land useless to mother nature and natural seed crops.

How do Monsanto use the poisoned land? The GENETICALLY MODIFY their patented agricultural crops to withstand this poison, so noting else can grow.

These genetically modified seeds are sold to desperate farmers, whom are not allowed to save seed for next year, and the farmers cannot collect viable seeds from their own crop!

WHERE DOES THE GENETICALLY MODIFIED POLLEN GO? – from the GM crops? It goes into the natural environment and introduces this trait to nature. This should not be happening.

SO their crops are produced to feed us? That is good right? ROUND UP kills everything biological on the field. SO WHY WOULD YOU WANT TO EAT IT? How often is your food cleaned properly?

WHY has there been ££billions spent on hiding the fact certain food products come from GM sources? If it is that good SURELY we should have the choice to buy it?!

LABEL GENETICALLY MODIFIED FOODS – at the very least

MONSANTO NEED TO TAKE RESPONSIBILITY FOR THEIR IMPACT ON NATURE

http://www.filmsforaction.org/walloffilms/

search for films on Monsanto, seeds and agriculture to understand the big picture.

Index

ActionAid 55, 188
activism 67, 109, 110, 125, 154, 159, 175, 183; alterglobalist activism 98; anti-GM activism 136; chef activism 109; civic activism 108, 110, 137; international activism 2, 45, 48, 171; NGO activism 179; political activism 144, 174; social activism 27, 50, 82, 97, 122, 187
advocacy: advocacy campaign 88; advocacy groups 53; advocacy science 13, 138, 141, 143, 174, 180–1, 183–4, 193
Africa 8, 9, 10, 12, 25, 39–40, 41–4, 47, 49, 50–5, 61, 80, 87, 143, 158, 164, 172, 174–7, 191
African Agricultural Technology Foundation (AATF) 54, 159, 176
AgBioWorld Foundation 91, 189
AGRA (Alliance for a Green Revolution in Africa) 56, 175
AGRAWatch 56, 188
agri-corporation 10
Agricultural and Rural Convention (ARC2020) 64
agriculture 6, 9, 80, 88, 90, 91, 99, 100, 103, 110, 115, 144, 180, 191; Biodynamic agriculture 21–7, 190; colonial agriculture 40–1; commercial agriculture 9; conventional agriculture 87, 157, 187; GM or transgenic agriculture 27, 52–3, 63, 68, 105, 123–4, 147, 156, 160, 164, 171, 175; industrial agriculture 100, 104, 108; intensive agriculture 106, 123; local agriculture 107–8; modern agriculture 47, 187; organic agriculture 19, 26–7, 47, 49, 72, 84, 86–7, 137, 158, 178–80, 190; small scale agriculture 99, 100; subsistence agriculture 48; sustainable agriculture 8, 91, 104, 106, 164, 180;

traditional agriculture 40–1, 51, 107, 152; transgenic agriculture 52–3, 68, 70, 105, 123–4, 160, 164
agrobusiness 8
Albania 66
alfalfa 9
allerginicity 82
Alliance for Food Sovereignty in Africa (AFSA) 55, 56, 177, 188
alterglobalism 12, 96, 105, 187, 192
Ammann, K. 82, 84, 86, 91, 92, 183
antibiotic resistance 82
Antoniou, M. 10, 81–4, 136, 143
Apple 160
apple 9
ARCI 110
Argentina 9, 27, 66, 166, 174
Asia 39, 40, 51, 61, 98, 99, 143, 177, 187, 10, 191
ATTAC 103
Australia 9, 25, 54, 56, 120
Austria 23, 63, 65, 66, 67, 141, 173
authenticity 107–8
autopoiesis 30
ayahuasquero 32

Bacigalupi, P. 124
Bacillus thuringiensis (Bt) 10, 83–8, 92, 157, 159, 162–5
Bacon, F. 52, 121
banana 53–6
Bangladesh 9
Baulcombe, D. 134, 153
Bayer 82
bean 9
Beck, U. 35
Bello, W. 98–102
bentgrass 9
Beobachter Initiative 33–5, 188